Your Brain
The Missing Manual®

Your Brain: The Missing Manual

BY MATTHEW MACDONALD

Published by O'Reilly Media, Inc., 1005 Gravenstein Highway North, Sebastopol, CA 95472.

O'Reilly books may be purchased for educational, business, or sales promotional use. Online editions are also available for most titles (*safari.oreilly.com*). For more information, contact our corporate/institutional sales department: 800.998.9938 or corporate@*oreilly.com*.

Editor: Peter Meyers

Production Editor: Nellie McKesson

Copy Editor: Alison O'Byrne

Indexer: Julie Hawks

Cover Designer: Karen Montgomery

Interior Designer: Ron Bilodeau

Print History:

        May 2008:   First Edition.

ISBN: 978-0-596-51778-6

[F]

Contents

Part 1: Warming Up

Chapter 1
A Lap Around the Brain 7

Chapter 2
Brain Food: Healthy Eating 27

Chapter 3
Sleep: Taking Your Brain Offline 45

Part 2: Exploring Your Brain

Part 3: Understanding Other People's Brains

The Missing Credits

About the Author

 Matthew MacDonald is an author and programmer extraordinaire. He's the author of *Excel 2007: The Missing Manual, Access 2007: The Missing Manual, Creating Web Sites: The Missing Manual,* and over a dozen books about programming with the Microsoft .NET Framework. In a dimly remembered past life, he studied English literature and theoretical physics.

About the Creative Team

Peter Meyers (editor) is the managing editor of O'Reilly Media's Missing Manual series. He lives with his wife, daughter, and cats in New York City. Email: *meyers@oreilly.com*.

Nellie McKesson (production editor) is a graduate of St. John's College in Santa Fe, New Mexico. She currently lives in Jamaica Plain, Mass., and spends her spare time making t-shirts for her friends to wear (*mattsaundersbynellie.etsy.com*). Email: *nellie@oreilly.com*.

Alison O'Byrne (copy editor) is a freelance editor from Dublin, Ireland. Alison has provided editorial services for corporate and government clients at home and internationally for over six years. Email: *alison@alhaus.com*. Web: *www.alhaus.com*.

Julie Hawks (indexer) has degrees in mathematics and library and information science. Much of her spare time is spent reading authors such as David Bohm, Ramana Maharshi, and Sri Nisargadatta as well as dreaming of traveling extensively through India. Email: *juliehawks@gmail.com*.

Esther Chung (technical reviewer) is a student of the brain at Wellesley College. She wishes to thank her good friend Shane Warden for his introduction, and Dawn Frausto for all her help during the tech review process.

Timo Hannay (technical reviewer) is a director at Nature Publishing Group, creators of *Nature* and other scientific journals, as well as a variety of online scientific resources. Among other things, he is the publisher of *Nature.com*, a co-organiser of Science Foo Camp, and a fluent Japanese speaker. He originally trained as a neurophysiologist at the University of Oxford, and as a biochemist at Imperial College London. Email: *timo@hannay.net*.

Jennifer Mangels (technical reviewer) is an associate professor of cognitive neuroscience at Baruch College, a senior college of the City University of New York, where she is principle investigator of the Dynamic Learning Lab (*www.baruch.cuny.edu/faculty/jmangels*). She also serves as Chief Research Officer for Lucid Systems Inc., a new company leveraging neuroscience methods in the domain of market research (*www.lucidsystems.com*). Because she has so much free time left over, she busies herself playing Balinese gamelan music.

Acknowledgements

This is the part of the book where the author is supposed to tell you that nothing would have ever been accomplished without the contribution of hundreds of impressively credentialed people who did all the real work. Well, allow me to depart from the script, because I could have done everything myself. However, the resulting book would have been short, incoherent, and hand-written on the back side of a paper towel roll. Fortunately, you don't have to read *that* book. Instead, you can enjoy a book that's been cleaned up, illustrated, and reviewed by some very sharp pencils. Best of all, it's been copied off the paper towels. In other words, if you enjoy your reading experience, you have the following people to thank.

First up are my big-brained reviewers, who contributed helpful insight and plenty of trivia. They include Esther Chung, Jennifer Mangels, and Timo Hannay, whose fascinating tidbit about the birthing practices of the hyena ranks as the most interesting piece of information you won't get to read about in this book. (You can get the exquisitely painful story at *http://en.wikipedia.org/wiki/Spotted_Hyena*.) Curiously, Timo was not the only reviewer to bring up the reproductive life of the hyena while reading this book. This suggests something deep and profound about the connection between cutting-edge neuroscience and randy animals, but I'm at a loss to say exactly what it is.

Second, I thank my editor Peter Meyers, who helped to indulge all my authorly desires (new sidebars, color pictures, fancy figures, you get the picture), and the supremely talented Robert Romano, who created the illustrations for this book. I also owe much gratitude to Akiyoshi Kitaoka, who graciously allowed us to use his rotating snakes illusion (page 12), Rhon Rorter, who created a few images that were adapted for the figures in this book, Nellie McKesson, who shepherded the book through its final stages, and the many people who worked to get this book formatted, indexed, and printed.

Lastly, I thank my family—particularly my parents, who lost many a neuron in their parenting years, and my wife's parents, who didn't fare much better. (In Chapter 10 they can all find out what went wrong.) Finally, I'm eternally grateful for my wife Faria and my daughter Maya, whose brains delight me in quite different ways, and I promise not to hook either of them up to an MRI machine to find out why.

—*Matthew MacDonald*

The Missing Manual Series

Missing Manuals are witty, superbly written guides to computer products that don't come with printed manuals (which is just about all of them). Each book features a handcrafted index; cross-references to specific pages (not just chapters); and RepKover, a detached-spine binding that lets the book lie perfectly flat without the assistance of weights or cinder blocks.

Recent and upcoming titles include:

Access 2007: The Missing Manual by Matthew MacDonald

AppleScript: The Missing Manual by Adam Goldstein

AppleWorks 6: The Missing Manual by Jim Elferdink and David Reynolds

CSS: The Missing Manual by David Sawyer McFarland

Creating Web Sites: The Missing Manual by Matthew MacDonald

Dreamweaver 8: The Missing Manual by David Sawyer McFarland

Dreamweaver CS3: The Missing Manual by David Sawyer McFarland

eBay: The Missing Manual by Nancy Conner

Excel 2003: The Missing Manual by Matthew MacDonald

Excel 2007: The Missing Manual by Matthew MacDonald

Facebook: The Missing Manual by E. A. Vander Veer

FileMaker Pro 9: The Missing Manual by Geoff Coffey and Susan Prosser

Flash 8: The Missing Manual by E.A. Vander Veer

Flash CS3: The Missing Manual by E.A. Vander Veer and Chris Grover

FrontPage 2003: The Missing Manual by Jessica Mantaro

Google: The Missing Manual, Second Edition by Sarah Milstein, J.D. Biersdorfer, and Matthew MacDonald

Google Apps: The Missing Manual by Nancy Conner

The Internet: The Missing Manual by David Pogue and J.D. Biersdorfer

iMovie '08 & iDVD: The Missing Manual by David Pogue

iPhone: The Missing Manual by David Pogue

iPhoto '08: The Missing Manual by David Pogue and Derrick Story

iPod: The Missing Manual, Sixth Edition by J.D. Biersdorfer

JavaScript: The Missing Manual by David Sawyer McFarland

Mac OS X: The Missing Manual, Leopard Edition by David Pogue

Microsoft Project 2007: The Missing Manual by Bonnie Biafore

Office 2004 for Macintosh: The Missing Manual by Mark H. Walker and Franklin Tessler

Office 2007: The Missing Manual by Chris Grover, Matthew MacDonald, and E.A. Vander Veer

Office 2008 for Macintosh: The Missing Manual by Jim Elferdink

Photoshop Elements 6: The Missing Manual by Barbara Brundage

Photoshop Elements 6 for Mac: The Missing Manual by Barbara Brundage

PowerPoint 2007: The Missing Manual by E.A. Vander Veer

QuickBase: The Missing Manual by Nancy Conner

QuickBooks 2008: The Missing Manual by Bonnie Biafore

Quicken 2008: The Missing Manual by Bonnie Biafore

Switching to the Mac: The Missing Manual, Leopard Edition by David Pogue

Wikipedia: The Missing Manual by John Broughton

Windows XP Home Edition: The Missing Manual, Second Edition by David Pogue

Windows XP Pro: The Missing Manual, Second Edition by David Pogue, Craig Zacker, and Linda Zacker

Windows Vista: The Missing Manual by David Pogue

Word 2007: The Missing Manual by Chris Grover

Introduction

This is a book about that wet mass of crumpled-up cell tissue called the brain, and why it's responsible for everything from true love to getting you out of bed in the morning. It's a book about how we think, how we feel, and why it's so difficult to stay away from that second piece of triple-chocolate cheesecake. It's a book that teaches you how to get a good nap, warns you never to trust a memory, and explains why—as successful as you may be—you'll probably never be much happier than you are right now (see Chapter 6).

There are many excellent books that explore the brain's anatomical inner workings. But in this book, biology takes a back seat to practical advice. In other words, you won't just focus on *how* the brain works, but on *how you can use it* more effectively. After all, your brain is easily your most important possession (or arguably a strong second place after that slick iPhone or those hot new shoes). It deserves proper upkeep.

Learning how to use your brain means delving into its quirks—and as you'll see in this book, the brain is *full* of quirks. Unwritten rules shape how the brain interprets a scene, reconstructs a memory, and solves a problem. Most of the time, these rules work to your advantage. After all, your brain is easily your most important possession; it deserves proper upkeep..

In this book, you'll learn to work around some of your brain's limitations. You'll also learn to enjoy the quirks you *can't* change (some of which make for great party tricks). Either way, by the time you've finished the last chapter, you'll have an entirely new understanding of the cauliflower-shaped organ that rules your life.

About This Book

This book is intended to be a practical look at how to get the most out of your brain. What makes it different from the average self-help guide is the fact that it's grounded in modern-day neuroscience.

This book has one advantage over most other books in the *Missing Manuals* series, which focus on computer software and cool gadgets: Unlike the headline-grabbing products of the high-tech world, your brain won't become obsolete anytime soon. Despite its weaknesses (such as slow calculation speed), its limitations (the need to spend a third of the day deeply asleep), its arguable bugs (optical illusions and nightmares), and its missing features (an auditory lie detector, emotion override switch, memory downloader, and so on), you're unlikely to find a way to significantly upgrade your brain in your lifetime. Microsoft will not release a patch. Apple will not supply a replacement. The only thing that will change is our understanding of what happens in the 100 billion neurons inside your cranium.

Learning how to use your brain often involves learning a bit more about its plumbing. For example, in this book you'll learn about the electrical pulses and chemical messengers that govern your thoughts, drives, and emotions. In these cases, a dash of neuroscience can explain quite a few of the stranger details of day-to-day life. However, there are whole fields of brain science that you won't touch, simply because they won't help you fine-tune your noggin. For example, you won't explore rare brain injuries or diseases that lead to puzzling symptoms, such as those popularized in Oliver Sacks' *The Man Who Mistook His Wife For A Hat* (Summit Books, 1985) or V. S. Ramachandran's *Phantoms in the Brain* (William Morrow, 1998)—both of which make for fascinating follow-up reading. Instead, you'll concentrate on tweaking and tuning your own mental hardware.

About the Outline

Your journey through the brain wends its way through 10 chapters:

- **A Lap Around the Brain (Chapter 1)** starts you off with a tour of the brain's biological machinery. You'll learn how the brain works, how it's evolved, and why.

- **Brain Food: Healthy Eating (Chapter 2)** goes in search of a brain-friendly diet. Along the way, you'll learn why your brain is an energy hog, how it uses cravings to get you snacking, and how you can fight inappropriate food urges.

- **Sleep: Taking Your Brain Offline (Chapter 3)** explores the question "What happens during your nightly 8 hours of oblivion?" (And if you've already answered "chase vampires, drive through tunnels, and appear at formal occasions wearing surprisingly little amounts of clothing," you'll be happy to hear that this chapter also attempts to nail down the riddle of dreams.)

- **Perception (Chapter 4)** leads you through the mirror-lined funhouse of perception. You'll look at brain-bending optical illusions, and see how the brain uses ingrained assumptions to lead you astray—over and over again.

- **Memory (Chapter 5)** explores the mysterious power of the brain to encode your skills and experiences. You'll meet two men who can't remember, and one who's cursed with never forgetting.

- **Emotions (Chapter 6)** enters the hot-headed world of feelings. You'll see how the brain processes fear and pleasure, and how it ratchets down happiness to prevent you from getting too comfortable.

- **Reason (Chapter 7)** explores how the brain reasons or, more frequently, how it avoids thinking with snap judgments and sloppy logic. You'll learn how to defend yourself against an assortment of common fallacies, and how to use creative thinking to solve vexing problems.

- **Your Personality (Chapter 8)** examines what makes you, well, *you*. You'll learn how scientists measure personality using five core factors, and get ready to take a test that exposes your own inner nature.

- **The Battle of the Sexes (Chapter 9)** goes where few dare set foot. You'll see how hormones sculpt the brains of men and women, and you'll consider how these chemical forces may (or may not) account for gender differences. Finally, you'll watch the fireworks happening inside a brain that's in love.

- **The Developing Brain (Chapter 10)** ends the book by looking at the timeline of your brain, from the moment it first developed in the womb to the time it will finally shudder to a halt. Along the way, you'll search for the cause of teenage moodiness.

Separating Truth from Speculation

Neuroscience evolves rapidly, and the insights in this book are based on its most recent discoveries. However, as with all scientific knowledge, there's always the possibility that better, more comprehensive studies will overturn the concepts we use today or change the way we think about them. In fact, it's a given.

When dealing with cutting-edge research, we've chosen not to bury you in footnotes. (Our basic feeling is that footnotes are only as good as the research on which they're based, and it's easy to cite a great deal of nonsense written by a great many people.) Instead, look to the language of this book to distinguish rock-solid truths from tantalizing speculations. When this book says "some scientists believe," you're about to meet a promising new idea that has some heavyweight neuroscientists behind it, but hasn't convinced everyone. When this book says "one study found," you're looking at some provocative new evidence that's on the cutting edge of brain research.

About MissingManuals.com

At the *www.missingmanuals.com* Web site, you'll find articles, tips, and updates to this book. Click the "Missing CD" link, and then click this book's title to see a neat chapter-by-chapter list of all the Web sites mentioned in these pages. You'll also find a brief bibliography with the books and Web sites referenced in these pages, and a few suggestions for further reading.

You're invited and encouraged to submit corrections and updates for this book. In an effort to keep it as up-to-date and accurate as possible, each time we print more copies we'll make any confirmed corrections you've suggested. We'll also note such changes on the Web site, so you can mark important corrections in your own copy, if you like. (Click the book's name, and then click the "View/Submit Errata" link to see the changes.)

In the meantime, we'd love to hear your suggestions for new titles in the Missing Manual line. There's a place for that on the Web site too, as well as a place to sign up for a free newsletter about the series.

While you're online, you can also register this book at *www.oreilly.com* (you can jump directly to the registration page by going here: *http://tinyurl. com/yo82k3*). Registering means we can send you updates about this book, including any additions or Web-only offerings.

Safari® Books Online

 When you see a Safari® Books Online icon on the cover of your favorite technology book, it means the book is available online through the O'Reilly Network Safari Bookshelf.

Safari offers a solution that's better than e-Books. It's a virtual library that lets you easily search thousands of top tech books, cut and paste code samples, download chapters, and find quick answers when you need the most accurate, current information. Try it free at *http://safari.oreilly.com*.

1 A Lap Around the Brain

For most of this book, you'll focus on what your brain *does*, and pay less attention to its plumbing. It's not that the brain lacks interesting hardware. But you can easily spend a lifetime studying your brain's biological workings without having the faintest idea why your company laid you off, your spouse ran off with another lover, and your dreams are filled with gorillas in tuxedos serving you shrimp cocktails.

To get practical information that can help with life's day-to-day challenges, you need to concentrate on your brain's *software*—in other words, the thoughts, emotions, and higher-level processes that are endlessly at work in your squishy gray matter. In this book, you'll explore these phenomena closely. But, before you get started, there are a few underlying details to get out of the way. You need a crash-course in brain basics.

In this chapter, you'll take a quick tour to see what your brain looks like and how it's structured. You'll take a close look at *neurons*—the tiny wires that convey electrical signals in your brain—and find out how your brain plugs into the rest of your body. Along the way, you'll dispel a few myths about the brain, peer into its evolutionary history, and learn a few of the secrets of mental health.

A First Look at Your Brain

It's time to meet your brain.

Lurking in the space between your ears is a very soft, reddish, jelly-like organ. (If you were expecting your brain to be firm and deep grey, like a wrinkled walnut, you are no doubt thinking of a *preserved* brain. The living brain is much squishier, and it's covered in deep red arteries.)

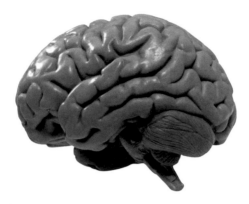

The average human brain weighs in at about three pounds. By comparison, an elephant's brain tips the scale at 11 pounds while a cat's brain—brace yourself, cat lovers—is a mere ounce. Bigger animals tend to have bigger brains, and some scientists suggest that a high brain-to-body weight ratio distinguishes the smart species from the dullards. In other words, the larger the brain is as a percentage of body weight, the smarter the creature. This calculation puts a few of our favorite animals at the top of the list (like dolphins and chimpanzees), but it needs a bit of fudgery to deal with really small animals (like birds and mice), which would otherwise appear to be raging geniuses.

 Tip You can check out the brain weight of your favorite animal at *http://faculty. washington.edu/chudler/facts.html*.

Of course, size isn't everything. Although all mammals have some strikingly similar brain hardware (and, to a lesser extent, so do all creatures that have any sort of brain), there are key anatomical differences. To really understand your brain, you need to dig deeper.

Dog
Brain Weight: 0.072 kg
Body Weight: 10 kg
Percentage: 0.72%

Cat
Brain Weight: 0.03 kg
Body Weight: 4.5 kg
Percentage: 0.67%

Rabbit
Brain Weight: 0.012 kg
Body Weight: 2.5 kg
Percentage: 0.48%

Sperm Whale
Brain Weight: 7 kg
Body Weight: 35,000 kg
Percentage: 0.02%

Nora
Brain Weight: 1.4 kg
Body Weight: 60 kg
Percentage: 2.3%

The Brain: An Archeological Site

Much as archaeologists examining an ancient site often find the ruins of multiple cities, each built on top of the previous one, neuroscientists peering into the brain find newer biological hardware built over the old stuff. In this section, you'll get the chance to peel back the layers.

The human brain is, like all the products of evolution, a work-in-progress. Although we won't see the human brain change in our lifetimes, millions of years of evolution have left their fingerprints all over it. Here's what's been happening:

- **The human brain has grown, becoming physically larger.** In fact, there's a strong case that humans suffer far more pain giving birth than almost any other animal because of our comparatively huge heads, which we need to carry around our outsized brains.

- **Existing brain hardware has been adapted for different uses.** The human brain is remarkably flexible. In deaf children, it can assign brain parts normally used for hearing to other tasks, like understanding sign language. In blind children, the brain can recruit the speech processing regions to interpret the tactile sensation of Braille letters. Over millions of years, similar but more profound shifts can occur. For example, many researchers believe that human speech hijacked some serious brain space in our early ancestors, and crowded out other skills.

- **New features have been bolted on top of old ones.** It's much easier for evolution to change what's already there than create a whole new brain from scratch. That means there's some deep, dark animal ancestry in your brain. If evolution were a building contractor, you'd find it leaving a few frightening things in the basement.

In the following section, you'll slice open your brain (metaphorically speaking) and get a closer look.

 Note No one knows why big-brained humans won the evolutionary arms race. Although it's tempting to conclude that smarter humans could build better tools (and therefore catch more nutritious animals), the brain has a significant evolutionary disadvantage—it's a hugely expensive energy hog. One of the more likely explanations for our success is that bigger human brains helped us attract mates and negotiate sticky group dynamics. In other words, we're all the descendants of a few sexy nerds.

The Outer Wrapper

The outer layer of your brain is the *cerebral cortex*. It powers conscious perception, abstract reasoning, speech, and creativity. It also accounts for over two thirds of your brain's weight.

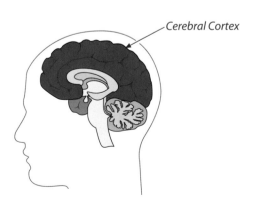

Cerebral Cortex

Although the cerebral cortex looks like a heavily wrinkled cauliflower, it's more like a crumpled sheet of paper. Its deep grooves and bulges allow the brain to cram in many more neurons than the less wrinkled brains you'll find in other animals. If you were able to stretch your cerebral cortex out flat on your lap, you'd find that it has about the same surface area as a page of newsprint from the *New York Times*. However, it's a bit thicker, and doesn't offer nearly as good a read.

 Note The previous picture shows the brain split down the middle. This view makes it clear that the cerebral cortex wraps along the top, front, and back of your brain. What the figure doesn't show is how the cerebral cortex also wraps around the sides of your brain. (To see the outside view of your brain, which is nearly all cerebral cortex, see page 8.)

The Middle Ground

Under the cerebral cortex, you'll find a set of older brain structures. These brain regions play a key role in memory (see Chapter 5) and emotional drives (such as the pleasure-seeking and pain-avoiding behavior you'll learn about in Chapter 6).

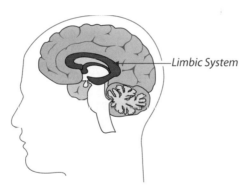

Limbic System

Sometimes, these brain structures are grouped together into a ring-shaped region called the *limbic system*. However, these days many neuroscientists doubt that they actually make up a distinct system that's separate from the rest of the brain. Instead, they prefer to examine each structure on its own merits.

The Basement

Buried deep inside your brain are its oldest structures, including the *brainstem* and the *cerebellum*.

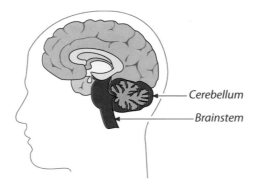

The brainstem looks like little more than a glorified lump at the top of the spinal cord. It controls body functions that have little conscious control, like breathing, hunger (see Chapter 2), and body temperature. It also plays the role of a massive conduit by funneling all the signals that travel between your brain and your body.

At the back of the brainstem is a fist-size growth that looks like a miniature brain. This region is the *cerebellum,* and it coordinates balance and movement. New research also suggests that the cerebellum plays a support role for other, more complex tasks. One theory is that it coordinates different regions of the brain so they can perform their work more efficiently.

The History of the Brain

It's common to describe the deepest regions of the brain as *older,* because these areas evolved first, in some distant species that was ancestor to us as well as many other modern creatures. For the same reason, we share these brain parts—in an extensively altered form—with other species. For example, bird and reptile brains appear to have a similar brainstem to ours but a vastly shrunken cerebral cortex. Orangutans, which are much closer relatives, have brains that are strikingly similar in almost all their component parts, but much smaller than ours.

It's a bit fanciful to imagine (as some early brain theories did) that the different layers of our brains are continuously at war. However, it's not so hard to picture a delicate balance between instinctive, ritualistic, and reactive behaviors that are rooted in the old brain systems and the morality, social sense, and problem solving that draw on the newer brain parts. In fact, it just might explain the paradox of a species that is equally at home in the symphony hall as it is on the field of war.

The Practical Side of Brain Science

Your Brain: The Ground Rules

Although it's sometimes difficult (and always controversial) to draw specific conclusions about human behavior based on brain anatomy, the evolutionary history of the brain suggests a few lessons:

- **Your brain is compartmentalized.** Different regions of your brain do different things. As you read through this book, you'll meet many of these specialized areas.

- **Your brain has competing systems.** Because the human brain was cobbled together over vast oceans of time, it's no surprise that its parts don't always work in harmony. For example, a sudden scare can cause your brain to briefly shut down its higher-level functioning and respond with the survival strategies that are coded at a deeper level (namely, "Run!"). This is one of the reasons you're likely to perform poorly at complex tasks when you're under stress (don't try to add complex sums when fleeing from a bear). You'll see a similar battle-in-the-brain when you consider how perception works with optical illusions (Chapter 4), or how logical thinking can overrule passionate emotions (Chapter 6).

- **Some things you can change; some you can't.** Some of the brain's most critical tasks are controlled by the most primitive areas of the brain, and they can't be overruled. For example, you'll have a hard time willing yourself to stop breathing, digest faster, or shift your internal body temperature a tenth of a degree, even though all these processes are controlled by the brain.

- **Your brain might not be suited for the modern world.** Reasonable estimates suggest that the last major change in the brain's anatomy dates back over 100,000 years. In other words, we're all living in a modern world with a somewhat outdated brain. How well the human brain has adapted to fast cars, fast food, and chronic stress is a matter of debate.

A Lap Around the Brain   **13**

Ancient Brain, Modern World

From an evolutionary standpoint, the human brain is a relatively recent development, with its sudden increase in size and pumped-up cerebral cortex happening just a few hundred thousand years ago. However, from the perspective of an individual human like yourself, the human brain is unimaginably old. This poses some sticky challenges, because the brain's survival strategies just aren't designed for 21st-century living.

This combination of old brain and new world hints at two of the key themes you'll explore throughout this book:

- **Your brain often works subconsciously.** As renowned neuroscientist Joseph LeDoux puts it, consciousness and language are "new kids on the evolutionary block." As the human brain evolved with its ever-expanding cerebral cortex, it became able to perceive, describe, and reflect on its own actions—many of which are unconscious and non-verbal. So don't be surprised when you find that your brain does many things without your consent, and many more without your realization. You may be able to understand what's taking place in the basement of your brain, but you can't always control it.

- **Your brain's logic doesn't always serve you well.** Every dieter knows that the brain's built-in circuitry can lead to trouble when confronting a larger-than-life billboard for the nearest fast-food chain. The problem here is that the brain has been honed by millions of years of evolution to be the perfect tool—for wandering groups of hunter-gatherers in the African savannah. For our ancestors, a good meal was hard to come by. But in the modern world where rich, nutrient dense foods are plentiful, the brain's natural response ("Eat Now!") can cause more harm than good. Similarly, it may be that certain brain disorders (say, obsessive-compulsive disorder) and some less-than-pleasant aspects of a properly functioning brain (like stress and nightmares) are the result of hardwired circuitry in older regions of the brain.

As you'll see in this book, your brain includes built-in circuitry that makes office politics seem like a life-or-death struggle (Chapter 6), tosses important facts out of your memory if they aren't charged with emotion (Chapter 5), and urges you to eat waistband-defying amounts of high-calorie snacks (Chapter 2). Sometimes, you can learn to compensate for your brain or work around its limitations. Other times, you'll be forced to accept its eccentricities.

 Note Evolution is a powerful brain-shaping force, but it's slow. It's a little bit like Microsoft asked you to create the world's most fantastic accounting software, you whipped it up, took a vacation for 100,000 years, and then came back with the package. Your program might still do the job, but it wouldn't be ideal.

The Brain's Wiring

So far, you've looked at the brain's shape, structure, and history. But you haven't yet seen it in action.

You probably already know that the brain is an electrical appliance more complex than any circuit board. But the brain also communicates with chemicals, using tiny compounds to transmit information, control mood, and interact with the rest of the body. Once you understand a few facts about your brain's wiring system, you'll have an easier time tackling some of the more sophisticated topics in this book.

Neurons

Your brain holds hundreds of billions of nerve cells. These cells come in two flavors: *neurons* (which get all the attention) and *glial cells* (which play an essential but often-overlooked supporting role).

Neurons carry electrical signals through your brain, and through the rest of your body. Estimates range, but the most widely cited calculations suggest that you have 100 billion neurons. (If you need an ego boost, compare that with the 300,000 neurons in the brain of the humble fruit fly.) Amazingly, there are at least 10 times as many glial cells, which provide nourishment, protection, waste disposal, speed enhancement (see page 228), and other support services for the spotlight-hogging neurons.

Here's a look at a single neuron:

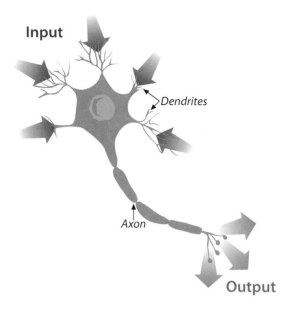

Up close and personal, a neuron looks like some form of futuristic vegetation. It receives messages through tree-like branches called **dendrites**. It then sends an electrical signal down a long tube-like structure called the **axon**. Add up the cumulative effect of several billion of these electrical impulses and you get a symphony, a treatise on law, or an episode of *Buffy the Vampire Slayer*.

 Note This picture of the neuron isn't proportionally accurate. In a real brain, the body of the cell (the top-left section of the picture) would be much smaller, while the dendrites, axons, and axon terminals (the branches at the end of the axon) would snake out far, far longer.

Synapses

The real magic happens when an electrical signal reaches the end of a neuron. At this point the neuron releases a bundle of chemicals into a tiny gap called a *synapse*. These chemicals, known as *neurotransmitters*, drift through the synapse (essentially "swimming" in the fluid of your brain) until they reach the dendrite of another neuron. That neuron can then react by firing its own electrical signal. In this way, a message can ricochet through the human brain, passing from one neuron to another.

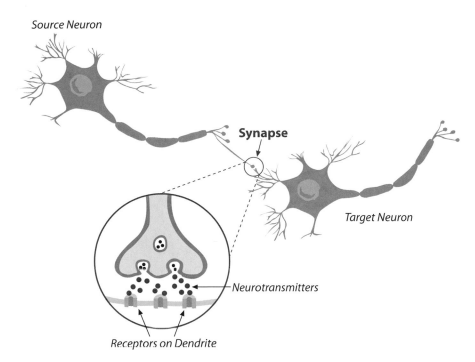

Source Neuron

Synapse

Target Neuron

Neurotransmitters

Receptors on Dendrite

As you might expect, this description is a huge simplification of the messy reality taking place inside your cranium. Here are some of the reasons why the brain's wiring system quickly becomes more complicated:

- **The brain uses different types of neurotransmitters, which affect different neurons.** Estimates suggest that the brain is a chemical soup, using more than 100 different substances to communicate between neurons.

- **An average neuron connects to several thousand *other* neurons.** That means thousands of neurons can simultaneously influence whether a single target neuron fires its signal. Similarly, one active neuron can pass its signal on to thousands more. It all adds up to a very flexible wiring system.

- **Neurotransmitters don't just trigger a neuron to fire.** They can also *inhibit* a neuron from firing.

- **Neurotransmitters don't just carry signals between neurons.** They can also act as *neuromodulators* to perform a host of different tasks. For example, a neuromodulator can alter the way a neuron works, change its sensitivity, trigger the creation of new proteins, and drift out of the tiny synaptic gap to affect entire regions of the brain. Many compounds in the brain act as ordinary message-carrying neurotransmitters in some situations, but behave like more powerful neuromodulators in others.

Note Neuromodulators may play a role in memory, learning, and mood control. For example, antidepressive drugs like Prozac work by increasing brain levels of serotonin, which can act as a neuromodulator. This change affects the way that billions of brain neurons work, in ways even the sharpest scientists don't currently understand.

If you could scoop out a small lump of your jelly-like brain matter and examine it under the microscope, you'd find a dense thicket consisting of millions of neurons, with dendrites and axons crossing and interweaving in an impossibly tangled fabric. It's estimated that the total number of connections between neurons (that is, the total number of synapses in the human brain) is in the eye-popping tens of *trillions*. It's for this reason that the human brain is sometimes described as the most complex object we've ever discovered in our universe. You should feel flattered.

Fun Facts

The Link Between Cosmetic Surgery and Sausages

Many of the world's most lethal poisons work by interfering with the way synapses work. One example is *botulinum toxin*, which is produced by a bacterium named *Clostridium botulinum*. Botulinum works by preventing the release of certain neurotransmitters. The result is that neurons can't communicate, and the brain rapidly loses the ability to send messages to the rest of your body.

Vanishingly small quantities of botulinum can cause death by paralysis (when eaten in an improperly canned tin of tuna), or remove wrinkles (when injected into the muscles of your face under the friendlier trade name Botox). Either way, botulinum toxin is one of the deadliest naturally occurring substances known to humankind. It's also the only neurotoxin named after a processed meat product. (*Botulus* is Latin for sausage, which can harbor fatal quantities of the botulinum toxin if it's improperly prepared.)

Poisons aren't the only substances that play with your neuro-transmitters. Many prescription drugs—and their shady black-market brothers—work by altering the chemical links that connect neurons.

The Nervous System

Often, we think of the human brain as a single device—a sort of biological computer made out of water, fat, and DNA. But the brain is actually a multi-pronged organ whose influence extends far beyond the head. In fact, the long tentacles of dendrites and axons stretch right out of the brain and into nearly every corner of the human body, uniting every muscle and organ into a body-wide network called the *nervous system*.

So far, you've learned how neurons can pass information between themselves. But the neurons on the outskirts of the nervous system get their input from something else. Depending on the type of neuron, they may fire signals in response to changes in heat, pressure (used for the sense of touch and sound), chemicals (for taste and smell), or light (for vision). These signals are then ferried up through the spinal cord to the brain. For example, a touch on your toe runs through just two giant neurons to reach your brain.

Similarly, an outgoing chain of neurons lets your brain send messages to the far corners of your body. When your brain needs to exert its control over a body part—either consciously or unconsciously—it simply triggers the right combination of neurons. The last neuron in the chain triggers the release of a chemical that kicks off the desired body process in another cell.

For example, if you stub your toe while line dancing, nearby neurons detect the deformation of your skin. These neurons pass the message up to the brain, which interprets this electrical activity as head-slapping agony. Your brain then triggers the neurons that will jerk the foot away. The last neurons in this sequence release a neurotransmitter to some nearby muscle tissue, compelling your muscles to contract and move your leg.

Of course, the low-level story is far more detailed. Even the simplest response involves many different neurons. For example, as you jerk your leg away using one group of muscles, you brain needs to relax another muscle group to prevent injury. Furthermore, the nervous system reacts to many different types of neurons in the same area of the body. This is one of the reasons that humans are "blessed" with so many types of pain. The dull ache of damaged tissue is picked up by a neuron that reacts to chemical changes, the flash of pain from a burn is triggered by neurons that react to extremely high heat, the sting of a cut is caused by neurons that react to the incision, and so on.

 Note The neurons that transmit different sensations take different pathways in the spinal column and sometimes travel with different speeds. Throbbing pain is the slowest, which is why you have a brief moment to contemplate the pain you're about to feel after smacking your toe on a door jamb.

The Endocrine System

As you've learned, your brain pulls all the strings. It controls a vast range of body processes simply by signaling the right neurons. However, neurons don't stretch everywhere, and they aren't nuanced enough to take every interaction into account. For that reason, your brain has another system that allows it to control the body—the *endocrine system*.

The endocrine system consists of a group of small organs known as *glands*. These glands work their magic by secreting various chemicals (called *hormones*) into your bloodstream. These hormones trigger reactions in other body parts. For example, the *thyroid gland* controls the speed of your metabolism. The *adrenal gland* controls the "fight or flight" response—it fires you up into a state of acute stress when an SUV steals the mall's last parking spot on Christmas Eve.

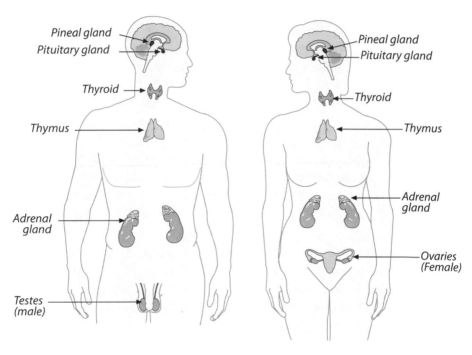

Much as the brain rules the nervous system, it's also the master of the endocrine system. Its key control mechanism is the *pituitary gland*.

The Pituitary Gland

To communicate with the glands in your body, the brain needs to release hormones into your blood. Its job is complicated by a defensive wall called the *blood-brain barrier*, which separates your brain from your bloodstream. The blood-brain barrier prevents most toxins, bacteria, viruses, and hormones from passing into the brain. The only substances that can pass through are extremely small ones, or ones that are soluble in fat. Fortunately, oxygen, alcohol, and caffeine make the cut. Other compounds need to be ferried across by specialized transporters. (One example is glucose, the sugar molecule that supplies energy to your brain.)

Much as the blood-brain barrier locks substances *out* of your brain, it also prevents substances from passing from your brain *into* your blood. To get around this limitation, the brain uses a built-in hormone dispenser called the *pituitary gland*. This pea-sized gland hangs out of the bottom of the brain, allowing it to slip hormones into your bloodstream whenever your brain gives the signal.

 Note The pituitary gland is often called the master gland, because it releases the hormones that tell the other glands (like the thyroid and adrenal glands) what to do. In this way, your brain can use the pituitary gland to exact precise control over the state of your body.

Although you may have vaguely heard about the pituitary gland before, it's already had a profound effect on your life. The brain uses the pituitary gland to release life-changing hormones at key points in your life. These hormones trigger growth and sexual development (as you'll see in Chapter 9), the contractions of birth, and milk production. Clearly, your brain is in charge of a lot more than you might have expected.

Incidentally, the part of the brain that controls the pituitary gland is the *hypothalamus*. You'll meet the hypothalamus several times in this book, starting in Chapter 2 (page 38).

Why Neurons Matter

Studying the anatomy of the brain is a great idea for socially starved medical students with plenty of free time. But even ordinary people can benefit from learning the nuts and bolts of neurons and synapses, because this knowledge provides the groundwork for understanding a variety of brain-related concepts.

A simple example is the process of sensitization and habituation. (*Sensitization* is why you're startled when someone drops a pen in a quiet study hall. *Habituation* is why you can still eat dinner while a construction company builds a new condo next door.) From studies on primitive-brained beasts, like giant squids, researchers know that these mechanisms are grounded in neurobiology—namely, the mechanisms that neurons use to close or open receptors to become more or less sensitive to neurotransmitters.

Here are a few more complex topics that will draw on your understanding of the brain's wiring:

- **Sleep.** In Chapter 3, you'll see how neuron activity in your brain changes when you're asleep and dreaming—and what it might mean.
- **Memory.** In Chapter 5, you'll consider how the brain doesn't file memories away in a separate storage tank—instead, it continuously rewires its structure by adding new synapses and removing ones that aren't needed.
- **Drive.** In Chapter 6, you'll see how the brain rewards itself by passing around the good stuff—specific neurotransmitters to the pleasure-obsessed neurons that crave them. (Incidentally, the same mechanism underlies many drug addictions. For example, opioids like heroin bind to specific receptors on neurons in the human brain. Ordinarily, these neurons are activated only at the brain's command, to cope with pain and to reward certain behavior. But by some cosmic coincidence, the poppy produces chemicals so similar to the brain's neurotransmitters that it can hijack key parts of our brain circuitry.)

Mental Fitness

You've now completed your first tour of the brain. Although you don't yet know all the reasons for the peculiar behavior of the planet's dominant species, you now have some of the tools that you can use to start asking the right questions. This makes it a good time to take a step back and change focus from low-level biology to more general guidelines. In this final section, you'll consider how you can keep your mental machine running in tip-top shape through the decades.

First, it's important to realize that the solution isn't to grow a bigger brain. After birth, it's rare for new neurons to appear in the brain. In fact, the story of the brain's development (which is told in Chapter 10) is largely the story of neurons and synapses dying off in waves as your body lumbers into old age. But don't panic yet. There's good reason to think that the loss of a few million neurons over the years is no big deal. In fact, it just might be part of the brain's natural housekeeping.

Rather than count the number of neurons in your head, it's more important to take note of the *connections* between them. As you've already learned, neurons are constantly being rewired. In healthy brains, the ratio of synapses to neurons grows as the number of neurons declines. In other words, leaner brains can become more efficient to compensate for their loss of neurons.

So what can you do to keep your brain in its best working form? There may be no way to dodge bad genes, bad luck, injury, and disease, but studies of brain aging consistently identify a few characteristics in old-aged but nimble-brained people. Here are a few practical guidelines if you hope to become a quick-witted fast-talking 90-year-old cribbage shark:

- **You are what you do.** The brain is constantly rewiring the connections between your neurons, strengthening the ones you use and weakening the ones that you don't. In other words, when you spend a day munching Cheetos, watching *American Idol* reruns, and lamenting the tragedy of your life, you aren't just whiling away the time. You're also training your brain to be a better Cheetos-eater, TV watcher, and chronic worrier. Fall into this pattern for a few years, and your brain just won't look the same.

- **Use it or lose it.** The brain may not be a muscle, but there's good evidence that the human body doesn't waste effort maintaining mental hardware that you never use. Surprisingly, it seems that it's never too late to ramp up your thinking. Many studies suggest that suddenly giving your brain more to do, even late in life, can overcome recent brain decline and foster broad, long-term improvements.

 Tip There's no magic brain-honing activity. But broad, integrative tasks like studying a new language, learning to play a musical instrument, changing jobs, writing a book, picking up a new hobby, and planning the perfect crime are all good prospects.

- **Embrace something different.** The brain craves novelty. The best way to keep your brain stimulated is to activate as much of it as often as you can. There's a fun side to this advice ("Indulge your curiosity!" "Engage strangers in long conversations!"), and a more challenging side ("Turn off the TV and learn differential calculus!"). The bottom line is that most of the time, the human body craves dull and easy stability. However, the brain *thrives* with constant challenges, tricky concepts, extreme concentration and, well, *work*.

 Note Don't think you can hone your brain with all-night Sudoku marathons. After the first 100 boards, your brain will have adapted itself to the patterns and strategies of the game, and will be able to polish off a board with far less neural work. On the one hand, this is a welcome development—after all, smart people use less activity for things they're good at. However, if your goal is to keep your brain strong, repeating the same type of challenge over and over again is no different than training with baby weights. For the maximum benefit, do something difficult and do something different.

- **Exercise the body to help the mind.** Studies suggest the keenest old brains have owners who exercise regularly. The best bet seems to be modest aerobic exercise, such as a daily jog or brisk walk. It's unclear why this helps, although it could well be that exercise stimulates other body processes that benefit the brain.

- **It wouldn't hurt to strum a tune.** The popular media is filled with tantalizing studies suggesting that a bit of music listening or music making can boost test scores and cultivate a baby genius. The truth is that the human brain is unlikely to respond to a magic music pill. However, exposing your brain to as many different influences as possible is always a surefire way to promote its development. Learning music as a discipline—in other words, as something to read, play, or improvise—is likely to draw on regions of the brain that are left dormant through the rest of your day-to-day life. (That said, if you're already an accomplished musician, your brain has long-ago transformed the challenging problems of making music into deeply ingrained neural patterns that take little effort. As a result, you'll get more brain stimulation by taking up accounting.)

- **Give your brain good food and rest.** Don't forget food and sleep. You'll learn all you need to know about brain-healthy eating and sleeping in the next two chapters.

Using 100 Percent of Your Brain

You may have heard the pernicious rumor that humans use a mere 10 percent of the brain's capacity. Not only is this statement utter hogwash, no one's quite sure who to blame for it. Although there are cases where people have done remarkably well after suffering damage to large portions of their brain (particularly if they were young when the injury happened), you'd be ill-advised to give up even a single square inch of your own brain tissue.

As you'll discover in Chapter 2, the brain is the greediest, most resource-sucking of human organs. Put simply, the human body would never pay the high price of keeping the brain online unless every single neuron increased its chance of survival.

2 Brain Food: Healthy Eating

S everal times a day, the average human puts whatever he or she is doing on hold and trundles off in search of some food.

At this precise moment, a small-scale drama unfolds in the brain. The deepest levels of the brain notice the shortage of food and trigger the physical feelings of hunger. The higher levels fire up food cravings, strategize about where to get the next food fix, and attempt to rationalize how a triple cheeseburger makes for a responsible breakfast. Here, the human brain shows its impressive abilities once again. In even the most well-adjusted person, it can convert a brightly-colored box of Oreos into a subtle interplay of desire, pleasure, guilt, and regret.

 Note It may well be that food guilt is the most reliable way to separate humans from other animals. Although other species have muscled their way into our territory in various other skill areas—demonstrating clear evidence of tool making, social bonding, and the contemplation of past and future—they aren't known to feel guilty after polishing off half a bag of ill-gotten dog food.

Clearly, the brain is deeply involved in the story of why (and what) we eat. In this chapter, you'll start by teasing apart the puzzle of food. For example, what does the brain do with all the calories it consumes? And how can you optimize its performance by eating the right foods? The answers aren't earth-shattering, but it all adds up to a good review if you don't have mom around to nag you about the virtues of a proper breakfast.

Next, you'll consider a subtly different side of the same issue—the human appetite. From a neurological standpoint, your desire for food is ruled by a cocktail of neurotransmitters and hormones that scientists have yet to puzzle out in its entirety. By exploring the biological basis for appetite, you'll gain insight into why many of us eat all wrong—and whether there's any hope to deny your brain the fast food, chocolate éclairs, and deep-fried Twinkies it craves.

The Brain's Energy Use

Your brain is an energy hog. Although it accounts for a fraction of your body weight (typically, about two percent), it devours an astounding 20 percent of the energy you use. And your brain's hunger is insatiable, whether you're asleep, awake, or focused on the very worst reality television. If your brain is deprived of energy for as little as 10 minutes, it suffers permanent damage. No other human organ is nearly as temperamental.

Before getting to the details of exactly how the brain gets its fuel, it's worth asking a preliminary question—namely, what the heck is the brain doing that it needs so much juice? Right now, your brain is using its calories in the following ways:

- Performing the normal housekeeping of all living cells, such as cleaning up debris, transporting nutrients, repairing cells, and so on.

- Building neurotransmitters (the chemicals that transmit messages from one neuron to another) and distributing them throughout your brain.

- Rewiring your brain circuitry with the new information you're learning.

- Firing electrical signals in your neurons, and keeping your brain's electrical system up and ready.

Out of all these tasks, the last one consumes the most energy. Neurons can easily fire an electrical signal hundreds of times a second, and a single neuron can talk to thousands of *other* neurons—each of which may also fire their own electrical signals to pass the message along. All this adds up to a lot of blinking lights in the big switchboard that we call the brain.

 Note Incidentally, your brain's energy use is roughly 20 watts—enough to power a very dim bulb.

Brain Fuel

Glucose—simple sugar—is the raw fuel that powers your brain. Unlike the muscles in your body, your brain can't tap the energy reserves in your body fat. (So thinking hard might tire you out, but it won't slim you down.)

Studies consistently find that very low glucose levels weaken the brain's ability to concentrate, remember, and pay attention. Some anthropologists even believe that our early ancestors kicked their brains into high gear when they discov-

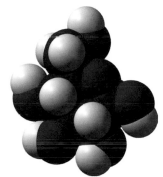

Your Friend: The Glucose Molecule

ered *starchy tubers*, a rich source of carbohydrates that can be readily broken down into sugar. (Potatoes, turnips, cassava and many other root vegetables fall into this category.) Although there's no concrete proof, the idea certainly gives French fry fans some serious food for thought.

Under normal conditions, the brain always gets the trickle of sugar it needs to stay functioning. However, certain drugs and diseases can bring on *hypoglycemia*, a condition in which even the brain's bare minimum sugar requirements can't be met. (For example, hypoglycemia is a possible side-effect of the blood-lowering medication taken by diabetes patients.) If this happens to you and your brain is deprived of sugar, you're likely to experience weakness, confusion, dizziness, and ultimately unconsciousness.

In other words, nothing craves glucose like a working brain.

Raw Sugar at Work

Now that you know your brain loves sugar, you might try to overclock it with a steady diet of chocolate, fudge icing, and gummy bears. Not so fast. The problem is that unlike the muscles in your body, your brain can only store the tiniest amount of glucose. Instead, it depends on your body to feed it a constant sugar supply through your blood. And simple, sugary foods don't stick around in your bloodstream for very long.

To understand the problem, consider what happens when you eat a quintuple-chocolate frosted donut:

❶ As your stomach digest the donut, your blood sugar rises. It's almost as fast as if you'd injected the sugar by syringe.

❷ Your pancreas (a small organ in your abdomen) notices the change and starts pumping insulin, which spreads throughout your body.

❸ Insulin tells the cells throughout your body to pull sugar out of your blood and store it for later use—except for cells in your brain, which lack the ability to keep a significant glucose stash.

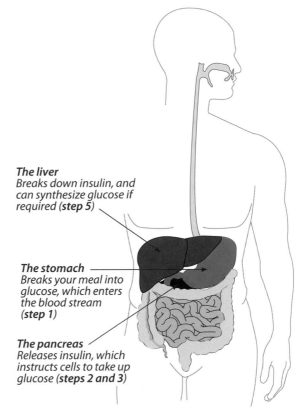

The liver
Breaks down insulin, and can synthesize glucose if required (step 5)

The stomach
Breaks your meal into glucose, which enters the blood stream (step 1)

The pancreas
Releases insulin, which instructs cells to take up glucose (steps 2 and 3)

❹ Your blood sugar drops to its normal level (thanks to the sugar-distributing power of insulin).

❺ Now, if your body and brain are hard at work, your blood sugar may drop even further. At this point, it's up to the liver to slip sugar back into your veins. Unfortunately, the liver works at a relatively slow speed. The muscles in your body have other ways of making do with a sugar drop (they can fall back on their own glucose reserves, or start chewing through fat). But your brain has no such help. The end result is that a short time later you may find yourself irritable, unfocused, and anxious—and in need of another donut fix.

The Sugar Rush Myth

For decades, popular wisdom has counseled you to avoid sugary snacks when clocking several hours of work, while recommending them for an emergency energy burst when wrestling with a short task. And most people, eager for an excuse to consume a few dozen Milky Ways when dire situations demand it, have swallowed this advice happily.

But don't unwrap that candy bar just yet. Recent studies suggest that the oft-described "sugar rush" may be little more than an urban legend. In average, non-diabetic people, the body is surprisingly efficient at monitoring the amount of sugar in the blood and quickly lowering it when it creeps too high. In fact, sugary snacks might just give your body *more* work to do, without giving your brain any boost at all. They're a good choice if your blood sugar does dip unusually low—for example, to revive yourself after extreme physical exertion—but not as a daily indulgence.

As you probably also know, a high-sugar diet is bad for another reason—it increases the risk of diabetes. Essentially, a high, see-sawing blood sugar level increases the likelihood that your pancreas will get tired of producing insulin and start slacking off, or your liver will get used to seeing insulin floating around and stop paying attention. Either way, the result is *type 2 diabetes*, a condition where the body can no longer manage its sugar supply effectively. Diabetes can cause trouble throughout the body and damage the brain. Although it's not known exactly how this brain damage takes place, diabetes sufferers are likely to suffer greater declines in brain functioning as they age.

 Note You don't need to worry about keeping consistent blood sugar. Your body does that automatically. However, if you strain your body's blood-sugar-stabilizing system a few thousand times too many, you risk the possibility that it may begin to fail, leading to diabetes.

Complex Carbohydrates: The Ultimate Time-Release Pill

To keep your brain running at peak performance for long periods of time, you can't mainline glucose with a Krispy Kreme donut. Instead, you need a regular supply of *complex carbohydrates*—the body's equivalent of a time-release sugar pill. Complex carbohydrates are bigger molecules that are broken down to simpler sugars over longer periods of time, keeping your blood stocked with a steady supply of the good stuff.

Complex carbohydrates are found in fruit, vegetables, nuts, seeds and grains.

 Note Protein—plentifully found in foods like fish, milk, and nuts—can also be broken down into glucose, but more slowly and with a bit more work from the body. Similarly, a small portion of a fat molecule can be converted into glucose, but most of it is used by other parts of the body, and is unavailable to the brain. Virtually no complex carbohydrates are found in meat, eggs, cheese, milk, or fruit juices.

All complex carbohydrates are not created equal. Different foods are broken down into sugar at different rates. Refined grains—found in nutritionally dubious foods like white bread and white rice—are quickly and easily converted to sugar. By comparison, starchy and fiber-rich foods take longer to digest, and superstar vegetables like broccoli, artichokes, and asparagus take even longer to yield their sugar reserves.

Responsible brain owners can use a ranking system called the *glycemic index* (GI) to pick out the foods that offer a slow and steady release of glucose. The glycemic index scores foods based on how quickly they give up their glucose payload to the body. Cheap-and-dirty high-GI foods provide a quick sugar fix, while healthier low-GI foods are converted into glucose slowly and over longer periods of time. The figure below compares the effect of a GI star (the eggplant) with that of a GI embarrassment (the bagel).

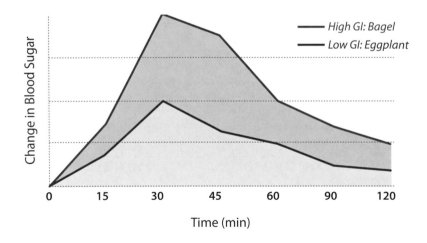

To get the profile on your favorite foods, check out one of the popular GI databases on the Web, such as *www.glycemicindex.com*.

Note Eating low-GI meals doesn't necessarily mean forsaking foods like white rice, bagels, and mashed potatoes. However, when eating these high-GI foods, be sure to pair them with some low-GI foods to bring the entire meal's GI to a respectable level, and ensure your brain keeps on an even keel until the next meal. Another good rule is to eat high-GI foods *after* the low-GI ones are already occupying your stomach. That's why dessert always follows dinner.

A Brain-Friendly Diet

So far, you've seen how your brain runs on glucose—and how to establish a steady supply. But although glucose is your brain's fuel, it's not the only ingredient your brain needs to stay shipshape.

Here are some other diet essentials for a balanced brain:

- **Protein.** Proteins are broken down into *amino acids*—incredibly versatile building blocks that the body uses to create a variety of compounds, including key neurotransmitters involved in attention and memory. This may be the reason that protein-rich meals appear to increase alertness (or this may just be a consequence of the fact that protein slows down the absorption of glucose, stabilizing blood sugar levels). Either way, it's a good idea to eat small amounts of low-fat protein at breakfast and lunch. Popular choices include yogurt, peanut butter, or a boiled egg. More exotic but equally nutritious choices include roasted crickets and steamed mealworms.

- **Fat.** Fat gets a bad rap, but it's actually responsible for a lot of essential functions in the body, and the brain is no different. In fact, your neurons are in large part built out of the stuff. Their membranes are composed of fatty acids, and their long axons are often wrapped in fatty insulation (which increases the speed that the electrical signal travels from one end to the other). However, not all fats are equal. Many studies suggest that the *omega 3* fats found in many fish are serious brain boosters. Diets rich in omega 3 fats are linked to healthier brains that have more resilient memories and a diminished risk of depression and degenerative diseases like Alzheimer's.

 Tip Although the exact benefits of omega 3 fats are still being debated, there's good reason to support the popular legend that seafood is brain food. Other omega 3 all-stars include avocados and olive oil.

- **Iron.** Iron plays a key role in transporting oxygen into the brain. Although there's no value in super-charging your iron intake, it's important to ensure you get a steady supply (through dietary sources such as red meat or vitamins).

- **Chocolate (and other antioxidants).** *Antioxidants* are a range of nutrients that disarm chemical troublemakers known as *free radicals* (which can damage cells and may underlie or aggravate cancer and a host of other health problems). The best way to get your antioxidants is to eat a range of fruits and vegetables, though cocoa-craving snackers can go straight to chocolate to get a boost of *flavanol*, an antioxidant thought to improve brain function. The only caveat is that the heavy-doses of sugar mixed with most chocolates probably mitigates the advantages of the flavanol. For best results, eat small quantities of the darkest chocolate you can find.

These aren't the only brain-altering substances that you'll find in your diet. The following items also turn up in many meals. However, they're potential troublemakers that you need to approach more cautiously.

- **Trans fats.** Trans fats are liquid oils that have been hardened into solids through a process called *hydrogenation*. The result is a cheap fat with a longer shelf life (and a few inconvenient health effects, like coronary heart disease). Trans fats are definitely not brain friendly. Studies suggest that when the body uses trans fats to create neurons, it creates feebler neurons that don't communicate as effectively. (Note to debaters and negotiators: Perhaps try harnessing this effect by supplying your opponents with complimentary potato chips?)

- **Caffeine.** You've no doubt encountered caffeine, the Western world's most popular stimulant. Caffeine is sold in various permutations, including as a blend of water and roasted fruit pits (where it's called coffee) and as a mix of water and dried leaves (where it's called tea). Either way, caffeine can stave off sleep and sharpen the brain's attention for short periods of time—as exam-cramming students, cross-country truck drivers, and unhinged comedy writers already know. (In mega-doses it can also cause headaches, abnormal heart rhythms, and anxiety.) The balance of current research suggests small quantities of caffeine are a safe indulgence. For best results, choose tea over coffee, which releases smaller amounts of caffeine over a longer period of time. Also, as the effects of caffeine vary from person to person, assess your own tolerance before you chug down that third double-espresso.

- **Alcohol.** Red wine is a cocktail of antioxidants. It may lower the risk of heart disease and it definitely plays a key role in making cringe-worthy family get-togethers more bearable. However, the effect of red wine and other alcohol in the brain is more controversial. Current science suggests that wine drinkers who regularly consume small amounts of alcohol with a meal (say, a glass of wine per day) are in the clear. But heavier drinkers don't get the same good news—they die with fewer neurons and shrunken brains.

Meal Sizes

At first glance, you might expect that large meals will give your brain a longer-lasting supply of energy. However, large quantities of food, particularly carbohydrates, trigger a chain reaction that prepares the body to rest and digest. It may not matter if you're lolling about on Thanksgiving, but if you need to concentrate on a difficult task it's best to avoid binging on that second plate of pasta.

As you'll learn later in this chapter, big meals knock your brain offline because of the brain's evolutionary programming, which is obsessed with food finding. An empty stomach secretes more of a hormone called *ghrelin*, which keeps you sharp, alert, and ready to forage for a meal. On the other hand, if your stomach is full, the brain concludes your caloric needs are temporarily satisfied. It's more likely to suggest that you relax, nap, and prepare yourself for the next snack attack.

Fun Facts

The Tale of the Turkey

The next time Aunt Ethel gives you the "lore of the turkey"—the oft-repeated Thanksgiving factoid that large quantities of turkey cause drowsiness—be prepared to set her straight. Although turkey does contain *tryptophan*, which is a building block for the neurotransmitters serotonin and melatonin (both of which can have a calming effect), the amount is small and has a hard time reaching the brain. In fact, turkey has no more tryptophan than chicken and beef, and less than cheddar cheese and soybeans. Rather, the turkey torpor of Thanksgiving is a simple result of wine, overeating, and Uncle Stan's scintillating conversation.

Meal Timing

Good eating is a mix of good food, low sugar, and solid timing. Studies have found that without a reasonable breakfast, production grinds to a halt in the brain. Children and adolescents are particularly at risk of spending the morning in a fog when they bypass the breakfast table.

The best advice is refreshingly obvious: Don't skip breakfast. Don't allow hours to pass without any food. Eat small meals and snacks throughout the day. However, you can safely put the stopwatch away. You don't need to schedule lunchtime to the minute, and you don't need to prefer frequent grazing over regular meals. For most people, three meals a day (with the occasional snack) will keep a healthy supply of fuel in the body's complex carbohydrate tank.

 Different people have different sensitivities to glucose. Studies have found that some people can gain a competitive edge from skipping breakfast before an interview or a nerve-wracking exam. (This edge stems from the attention-stimulating bite of hunger and stress hormones, which the brain uses to get you in prime food-hunting form.) But for others, missing breakfast is akin to running on a tank with only millimeters of gas left. It not only leads to fuzzy thinking in the morning, but also midday food binges and afternoon sluggishness.

The Practical Side of Brain Science

Three Simple Rules for Good Eating

In general, what's good for the body is good for the brain. The three best dietary guidelines you can follow are:

- Chose complex carbohydrates (like brown rice and whole-grain bread) over refined products (like fluorescent-colored pressure-molded breakfast cereals).
- Choose omega 3 fats over their more prevalent omega 6 fats. (Although both are important, the Western diet skews heavily in favor of the latter.) The best source is fish (particularly salmon, tuna, sardines, anchovies, herring, and mackerel).
- Limit the total amount of food you consume.

Here's a sample meal plan that keeps your brain in gear all day; for a print-able copy of this chart visit this book's "Missing CD" page at *www.missing-manuals.com*.

Meal	Recommended Food	Example
Breakfast	A modest helping of complex carbo-hydrates with a dash of protein.	Oatmeal and yogurt (of the low-fat, low-sugar variety) A fried egg with whole wheat toast
Late Morning Snack	More complex car-bohydrates. This is an ideal time to get those fruit and veggie servings in.	Fruit salad Carrot sticks
Lunch	Protein and complex carbohydrates.	A tuna sandwich with salad Salmon and squash over brown rice
Afternoon Snack	Another helping of complex carbohy-drates, with extra points for fruit and veggies.	Tomato salad Strawberries and cashews
Dinner	Complex carbohy-drates with protein and a dash more fat to get you through the night.	Roasted chicken and sweet potato Seafood paella
Evening Snack	A very small amount of sugar and fat helps put the brain to sleep. But be warned—des-sert is a rocky coast that dashes many good intentions!	Dark chocolate Tea and a cupcake

The Secret Gears of Appetite

You've now learned what the brain does with your dinner. However, you haven't considered *how* it gets what it wants—in other words, what neurological process underpins the hunger pangs that can drive you out of bed for a midnight snack or cripple your resolve when strolling past the vending machine.

In truth, the full appetite story is still shrouded in mystery. This isn't because the human appetite is a particularly strange phenomenon, but because there are many overlapping influences that come into play. At any given moment, your desire to eat (or ignore) food is shaped by the time of day, the current fullness of your stomach, your emotional state, and the amount of fat, sugar, and protein that's circulating in your body.

Although even the sharpest brain scientist can't discern the appetite's exact equation, we do know the brain center that evaluates these factors and triggers your hunger. It's the *hypothalamus*, the ancient control center that sits at the top of the brain stem. (You first met the hypothalamus in Chapter 1, where you learned how it controls the pituitary gland, the brain's 24-hour pharmacy shop.) In studies with unfortunate rats, scientists discovered that damage to one section of the hypothalamus causes rats to lose their appetite and willingly starve. Damage to another section causes rats to eat insatiably and balloon up to three times their normal size.

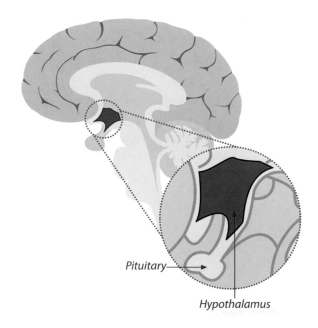

Pituitary

Hypothalamus

The appetite-controlling system of the hypothalamus is surprisingly complex. The hypothalamus includes neurons that react to the distension of the stomach, and others that respond to the levels of sugar and fat in the blood stream. It also pays attention to two more recently discovered hormones: ghrelin and leptin.

Ghrelin and Leptin

They may sound like two nasty hobbits, but these two hormones play a key role in shaping your appetite.

- **Ghrelin.** This hormone is produced by the lining of your stomach. Its presence rises before meals, and falls after you eat. Ghrelin appears to act on the hypothalamus to stimulate appetite. In studies, a quick shot of ghrelin gave participants a voracious appetite worthy of an all-you-can-eat Chinese buffet.

- **Leptin.** This hormone plays the opposite role. It's released by fat cells and acts on your brain to reduce appetite. Mutant mice that were bred *without* leptin receptors (in the neurons of their brains) ate more and grew to eye-popping sizes.

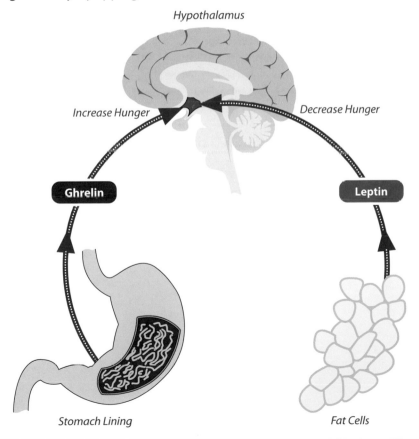

Based on this information, it's easy to conclude that your brain is actually under the control of your body, be it the ghrelin-releasing stomach lining or the leptin-releasing fat cells. But the truth is quite a bit stranger than that.

Recent studies suggest that leptin and ghrelin are part of the body's critical anti-starvation system. Supporting this theory are the fact the ghrelin levels are sky high in undernourished patients with anorexia and obese patients who have quickly slimmed down on a liquid diet. In both these cases, the body is detecting what it suspects is a critical shortage of calories (in the first instance correctly, in the second instance not), and signaling to the brain that it's time to down some serious chow. In other words, ghrelin and leptin may represent just *one* of the ways that the brain monitors the state of your body.

Here are two more factors that complicate the picture:

- Ongoing studies are continuously unveiling other chemicals that play a role in hunger and satiety (feeling full). A few additional players include galanin, enterostatin, and obestatin.

- Many chemicals that are involved in appetite also have other purposes, such as playing a role with growth, puberty, sleep, and glucose regulation.

It may well be that all these chemicals are part of a multilayered appetite-driving system that has a built-in backup. In other words, if ghrelin isn't enough to get you to chow down (or if some part of that exchange doesn't work correctly), other processes may pick up the slack. On the whole, your brain is blissfully unconcerned with overeating, but worries chronically about impending starvation.

 Note Lack of sleep appears to boost ghrelin levels and deplete leptin, which may be one reason why the chronically sleep-deprived have a greater incidence of obesity. So if you're trying to tame your waistline, try to start with a solid eight hours of snoozing.

Your Anti-Starvation System

Your brain's food-related behavior is another reflection of its evolutionary past. Over the millions of years of evolution that led to the human brain, it's unlikely that overeating claimed many lives. Instead, the most successful people were those who could eat gluttonously in times of plenty (and hold onto their spoils with fat-hoarding genes). Scarce food supplies killed off the picky eaters in droves.

The brain's evolutionary history also holds the clues that explain why we love the tastes we do—namely, sugar, fat, and salt. All three are essential for the human body, and all three were in scarce supply before the modern world. The most successful humans were those whose taste buds could spot these rare ingredients in the wild. And the most successful of *those* humans were the ones who could gorge themselves on belly-stretching quantities of food until the next unlucky antelope wandered past. It's no wonder that modern-day dieters feel the deck is stacked squarely against them. They, like the rest of the human race, are descended from a few early humans who had the genetic programming to seek out anything remotely resembling a Big Mac.

This all adds up to some depressing news for the chronic dieter. Not only is the brain preprogrammed to gorge itself on calories when the opportunity presents itself, it also craves the contemporary world's over-abundant junk foods. In a few more million years, evolution may straighten out the picture and we may all be replaced by a race of veggie-craving light eaters. In the meantime, we're forced to navigate the modern world with a brain that's designed for a distinctly different environment. It's a bit like buying the hottest possible racing car and plunking it down in the heart of Los Angeles traffic. There's a mismatch between the expectations of your hardware and the provisions of your environment, and the resulting experience isn't going to be pleasant.

Unfortunately, the story gets worse.

The Set Point Trap

The *set point* theory suggests that your human brain has an ideal weight picked out for you. If you slim down from this genetically programmed weight, the brain tweaks the gears and dials in your body to try to get back to it—for example, ramping up hunger while slowing down metabolism, and increasing ghrelin while lowering leptin. In essence, this is another side of the body's robust anti-starvation system.

The problem is that the body is much more willing to let the set point creep *up* than drift *down*. In other words, if your waistline expands, that's where your body and brain expect it to stay.

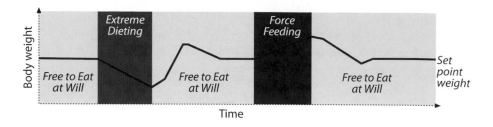

The set point theory is every dieter's nightmare. However, it's not an insurmountable challenge. After all, human bodies and brains are remarkably adaptable. If you're battling an unhealthily high body weight, the best advice is to proceed slowly, giving your brain ample time to adjust and establish a new set point. And if you're at a healthy weight, avoid the two things that are likely to force your set point up—extreme eating and insufficient exercise.

Emotional Eating

So far, you've focused on the biological triggers of overeating. As counterproductive as these triggers may be in the 21st century, they aren't enough to cause obesity. After all, even though the human brain is preprogrammed to love food, it isn't daft enough to pick a set point of 400 pounds. Instead, severe weight problems are often the result of a potent mix of human nature and biological bad luck.

Many overeaters battle the temptation to eat large amounts of sweet and fatty foods in emotionally trying times. Unfortunately, the brain has biological circuitry that reinforces this bad idea. As you learned earlier, a light meal keeps the brain alert and active, while large quantities of food put it into a more relaxed, lethargic state. Big portions of sweet and fatty foods also dampen stress hormones, cause the brain to release pain-relieving substances, and trigger the brain's reward circuitry (thereby congratulating the eater for a food-foraging job well done).

 Note The German language describes the phenomenon of emotional overeating with the evocative word *Kummerspeck*, which translates literally to "grief bacon."

It's easy to see how these mechanisms can contribute to the ever-expanding waistlines of the chronically stressed, depressed, and sleep-deprived. But studies show that even healthy individuals are compelled to eat more after watching a depressing movie, in an apparent attempt to sooth themselves. In fact, the magnitude of the effect is similar in self-confessed emotional eaters as in more cool-headed participants.

The Practical Side of Brain Science

Controlling Your Hypothalamus

Many people have no trouble customizing their diet to include appropriate quantities of wholesome, junk-free foods. But many more wrestle with fat, sugar, and overgrown portion sizes. You can understand these problems as a result of the brain's hard-wired genetic programming, but they aren't easy to overcome.

In a face-to-face match between the hypothalamus and you (that is, your higher-brain regions, which we assume are the ones reading this chapter), you can win the battle and overrule the demand to eat. However, you can only win temporarily. The hypothalamus will wage an unending war until you capitulate to its demands.

So faced with an opponent you can't beat, what can you do? The best advice is to plan ahead and change the rules—in other words, ensure that the great dinner battle takes place on your terms. Here's how:

- **Stack the deck.** If you can set up an environment that makes it easy to eat the right foods, you'll be less tempted to transgress. In other words, it's easier to deny yourself the fudge-encased ice cream sandwiches before they get into your shopping cart than it is once they're beckoning you from the freezer. Similarly, the best time to prepare a healthy meal is before you're crippled by pangs of hunger (and give in to a box of ready-to-eat glucose).

- **Involve your higher brain.** The worst food transgressions might actually happen while your brain is on autopilot. These habits—eating the first food you see, eating until the plate is empty, merrily shoveling in extra calories while you talk—are a part of hard-wired brain logic that you can easily beat, once you become conscious of it. For a great description of the problem (and some memorable experiments, including a soup bowl that surreptitiously refills itself), check out Brian Wansink's *Mindless Eating* (Bantam, 2006).

- **Eat like your ancestors.** It's easy to be paralyzed by food angst, especially in a modern supermarket where probiotic-fortified Twinkies promote themselves as a sensible breakfast. However, you can navigate food puzzles by choosing food your great grandmother would approve of. For most people, this single rule is enough to steer clear of processed foods, fad diets, and large quantities of fat and sugar. For more insight into the decline of sensible eating in modern society and a few more guidelines, read Michael Pollan's *In Defense of Food* (Penguin, 2008).

3 Sleep: Taking Your Brain Offline

Sleep is one of the quirkiest brain behaviors. If it wasn't such a fundamental part of your life, you'd find the whole idea more than a bit outlandish. Think about it: For nearly a third of the day, your brain paralyzes your body. It then slips into a state of supposed rest that has bursts of electrical activity as energetic as when you're awake. And to top things off, the sleeping brain reels with hallucinations that rival those induced by the most potent controlled substances.

Scientists who study sleeping brains have unearthed all kinds of fascinating things. But they still can't agree on why we do it. In fact, they still can't completely agree that we actually *need* to do it. And the story gets even stranger when neuroscience shifts its attention to the surreal world of dreams.

In this chapter, you'll take a long, sober look at the sleeping brain. First up: a consideration of possible reasons your brain craves sleep (including a look at why it entertains itself with wild, convoluted flights of fancy while you're out cold). As you size up the science of sleep, you'll also dip into its many practical uses—for example, how sleep bolsters learning, how to harness the creativity of your dreams, and how to get a good nap.

Your Biological Clock

Most humans are well adjusted to the basic schedule of modern life—sleeping through breakfast, dozing off after lunch, and watching late night television when they should be deep asleep. Against this backdrop, it's amazing to realize that every human has a built-in timepiece that, if properly calibrated, can get you to bed at night and up in the morning with flawless punctuality.

This time-keeping device is embedded in a region of the brain called the *suprachiasmatic nucleus* (SCN). This small bundle of neurons is a part of the *hypothalamus*, which—as you've discovered in previous chapters—is a deep, ancient structure in the core of the brain that performs key tasks, like regulating the release of hormones and controlling appetite.

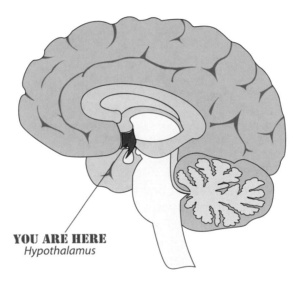

YOU ARE HERE
Hypothalamus

The Circadian Rhythm

Scientists have discovered how the SCN works by putting good-natured people in dark caves for long amounts of time. Not only is this an entertaining way for brain researchers to while away a weekend, it also turns out to be surprisingly informative. Confining people in caves tells us how humans manage their time when they have no external cues to indicate whether it's morning, midnight, or midday.

 Famous time-isolation studies have used actual caves, an underground glacier, a bomb shelter, and less impressive-sounding research laboratories.

During cave studies, volunteers are free to sleep whenever they like. However, they gravitate to a 24- to 25-hour cycle that closely resembles what we think of as a normal human day. As this cycle, called the *circadian rhythm*, draws to its close, the participants get ready to sleep. As the cycle starts again, they pass through their deepest sleep, and then rise to start a new day. The cave studies show that you don't need the rising and setting of the sun to know when to get out of bed. Instead, the SCN keeps an internal clock running all the time.

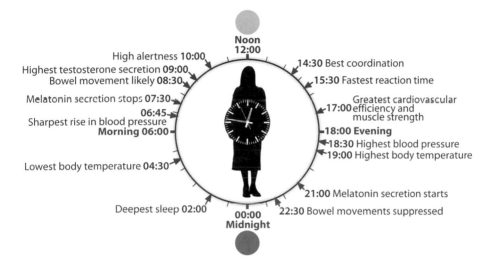

The circadian rhythm doesn't just govern sleep and wakefulness. It also influences a host of body processes that vary over the course of a day. For example, body temperature is at its lowest in the early morning, and it peaks in the evening. Similarly, rote memory (stuff you memorize by repeating over and over) is keenest before lunch, and coordination is best in the afternoon (around 2:00 p.m.). These daily schedules are controlled by an intricate family of hormones. These schedules also affect diseases and chronic conditions. For example, the early morning tends to be the most difficult time for people with rheumatoid arthritis and asthma sufferers (along with the late evening). It's also a risky time for heart attacks.

 Tip Understanding the circadian rhythm (and your own body's slight variances) can help you choose the best time to perform certain types of tasks, like studying, relaxing with a good book, or breaking out your hip hop dance moves.

It's interesting to ask whether the body's biological clock is set from birth, or whether the SCN starts off as a more flexible clock that learns the 24-hour day cycle from experience. Although proposals to place infants in caves haven't faired well, researchers have some insight from studies on other animals (namely fruit flies and rats). The current evidence suggests that the SCN includes specially adapted time-keeping cells that are hard-wired to expect a cycle of about 24 hours. Exposure to light can shift this schedule, but changes are slow and limited—hence our trouble moving from one time zone to another.

The Perfect Day: 24 or 25 Hours?

In many versions of the cave experiment, the cave dwellers unwittingly stick to a 25-hour schedule. As the study wears on, the cave "day" becomes more and more out of sync with the outside world. Ten days into the experiment, cave dwellers are heading to bed before lunch (real world time) and rising just after sunset.

There are two possible explanations for this divergence:

- **Humans prefer 25-hour days.** This would explain quite a few Saturday sleep-ins (and give you an ironclad excuse the next time you're late to the office). However, it seems oddly out of touch with the daily cycle of the planet Earth.

- **Artificial lights skew our sense of time.** Some theories suggest that the human clock is within minutes of 24 hours, but it's thrown off track by our ever-present artificial lights. Essentially, indoor lighting fools the SCN into thinking the sunset is later than it really is, thereby shifting our clocks forward. This explains why even the most sleep-deprived cubicle worker tends to stay up late in the evening, without feeling particularly tired until it's already too late to get a decent sleep.

 Note The SCN is wired into the optic nerves that connect to your eyes. This is how the SCN gets the signal that says it's daytime.

Why We Sleep

Now that you know about the sleep-regulating pacemaker implanted in your brain, you may be curious about *why* it exists. In other words, why is sleep so important that there's a module in your head dedicated to nagging you about it?

Getting a Good Sleep

As you've learned, there's a highly accurate clock lodged in your brain. This timepiece could do a splendid job running your life, getting you out of bed without an alarm clock and sending you to bed before sluggishness sets in—if you weren't scrambling its time-keeping with the bad habits of modern living (and possibly artificial lights).

Fortunately, now that you understand how the SCN works, you can improve the situation. Here are some good tips:

- **Use light to shape your sleep.** Having trouble sleeping on time? The easiest way to adjust your biological clock is with light. So turn it on bright to wake up in the morning and dim it to slow down in the evening. And if you really need a solid night's sleep, try to make do with no artificial lights at all (or shift to weaker sources of illumination, like candles, oil lamps, or a roaring fireplace). If all else fails, relocate to a darkened cave for the night. (If you've ever had your light constrained by the sun—for example, when camping or in the midst of a power outage—you've probably noticed how uncomplicated sleep becomes.)

- **Work late, but not all night.** If you're planning to work late, set 4:00 a.m. as your absolute limit. At that point, your body is ready to shut itself down for a couple of hours at least. Those who stay up past this point will have poor coordination, slow reflexes, and are at an increased risk for accidents.

- **Use the sun to bridge time zones.** If you're traveling across time zones, give the sun a chance to reset your biological clock. For example, a walk in the bright sun is a perfect way to adjust your clock on arrival. You can begin shifting your clock a day or two ahead of your trip. To do so, hunt down a jet lag calculator on the Web, or just keep two good rules of thumb in mind. Before traveling *westward*, expose yourself to light in the late afternoon and evening, and avoid light in the morning as much as possible. When traveling *eastward*, expose yourself to light in the morning, and avoid light in the evening as much as possible. In general, it's more difficult to adjust to eastward travel.

- **Avoid shift work.** If you have to do it, understand that it's a no-holds-barred body-defying challenge. To stay as healthy as possible, try to give your body the cues of a normal day. For example, sleep in a completely darkened room, and get bright full-spectrum lights to shine when you wake up. Follow the eating habits of a proper day, with breakfast after you rise, a decent lunch halfway through, and a light dinner a few hours before you prepare to sleep.

- **Don't drink and doze.** Alcohol can help you get to sleep, but it wreaks havoc with the sleep cycles you'll learn about later in this chapter, ultimately leading to a less restful night. If you do indulge in a late-night tipple, stay awake afterwards to give your body enough time to metabolize the alcohol (and down a few glasses of water to stave off late-night dehydration). The effect of another popular beverage, milk, is much more controversial. Science suggests it's not likely to have the sleep-inducing properties your mother claimed, but a warm mug in the evening certainly can't hurt (unless you're lactose intolerant).

At first glance, a bit of common sense seems like enough to answer this question. After all, lack of sleep makes most people cranky and foggy-headed. However, this state might simply be the work of the SCN and the circadian rhythm—in other words, you feel lousy when you don't sleep because your brain *wants* you to sleep. This still doesn't explain why your brain's so keen on conking out.

 Note Popular wisdom is that sleep is a restorative process—a chance for the body to repair itself and for the brain to relax. This is partly true. If you're healing from a wound, fighting off an illness, or recovering from extreme exercise, sleep speeds recovery. But otherwise, the picture isn't nearly as clear. Although the sleeping brain changes gears, it stays active for most of the night. And while it's possible that this shift gives the brain a chance to clean itself up, it's equally possible that sleep has another purpose.

Sleep in the Animal Kingdom

Comparing humans to other animals raises even more questions. Nearly every form of life follows some sort of 24-hour clock. Even animals without an SCN (like fruit flies) have cells that work in much the same way as the neurons in the SCN to keep time. In fact, even plants keep a 24-hour clock to control growth, reproduction, and leaf movement.

However, despite these astounding similarities, there's a huge variability in the amount of time different animals sleep. In general, predators like lions and tigers have the luxury of sleeping in (and both log daily sleep times of over a dozen hours). Cats, their supremely lazy descendants, follow suit. On the other hand, commonly sought-after prey (say, gazelles) get by with just a few hours a day, and usually take it in short bursts that last mere minutes. Humans fall in the middle.

There are two reasons why prey sleep for shorter periods of time. First, sleep is dangerous, given the many hungry animals around looking for a snack. Second, the diet of prey animals is typically limited to grasses and vegetation. Combine this low-calorie diet with the metabolic needs of being constantly alert and ready to run and you'll see why gazelles need to graze away almost all the hours of the day to stay alive. Predators, on the other hand, can gorge themselves on meat and then skip a few meals.

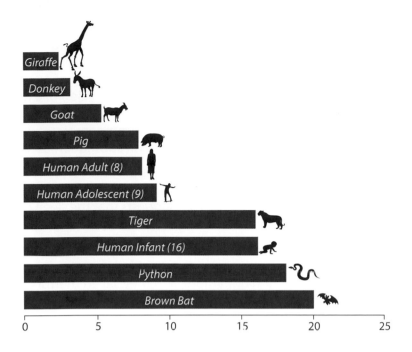

Giraffe					
Donkey					
Goat					
Pig					
Human Adult (8)					
Human Adolescent (9)					
Tiger					
Human Infant (16)					
Python					
Brown Bat					

0   5   10   15   20   25

 Note Not only do animals sleep for different amounts of time, but they also sleep In different ways. Extreme examples include hibernating animals, like bears, that doze an entire season to forego the inconvenience of roaming about. Dolphins are even stranger. They have the remarkable ability to keep an ever-watchful eye on their surroundings by putting just half of their brain to sleep at a time. (Humans still lack this ability, despite many of our best attempts.) Both these examples from the animal kingdom suggest that the natural requirements of sleep are a lot more flexible than one might assume.

Sleep in Humans

The sleep habits of other animals suggest a common theme—namely, animals sleep according to their lifestyle. The problem is that the human lifestyle has recently changed quite a bit. All this is enough to make chronic under-sleepers wonder if our eight-hour-a-day sleep schedule is just another inconvenient legacy from our long-lost ancestors.

As with most questions involving the brain, the answer is a bit murky. Some scientists (including leading sleep researcher Jerry Siegel), believe that human sleep is little more than a strategy for staying out of trouble for as long as possible—meaning we're wasting a lot of time lazing around in bed. One fact that lends support to this claim: animal species that have better access to safe shelter sleep more. In other words, animals sleep for as long as they feel they're safe.

Other scientists argue that sleep is critically important, or at least clearly worth the time. After all, the brain seems to crave it. People who miss a few hours every night build up a *sleep debt*—that is, they need to sleep longer to put their brain back on an even keel. (Later in this chapter, you'll find that this also applies to the different stages of the sleep cycle. For example, if people are allowed to sleep but are interrupted whenever they begin the vivid dream stage of REM sleep, their bodies will compensate by getting back to REM sleep faster and staying in it longer on subsequent sleeps.)

Furthermore, humans fall squarely between the predator and prey sleep ranges. We don't sleep as soundly as an antelope-stuffed lion, but we also aren't as restless as a paranoid gazelle. That suggests that the sleep cycle in humans might not be a slave to the biological rules of predator and prey.

Instead, it's quite possible that our distant ancestors gained a few modest benefits from night sleep—for example, they avoided bumping into things in the dark (the cost of poor night vision), and they saved themselves the energy cost of staying active in the colder nighttime temperatures. However, evolution is a relentlessly practical force that's always willing to capitalize on an existing behavior by reusing it for something else. So once the earliest ancestors of humans got used to their 8 hours of sleep, later, more modern humans might have started using it for something else—like dreaming.

Long-Term Sleep Deprivation

One way to explore the importance of sleep is to see what happens when people don't get it. Skimping on these eight-hour installments of unconsciousness (or skipping them altogether), causes short-term problems, like difficulty focusing and irritability. However, the next good sleep puts everything well again. People have clocked over 10 days without sleep. They've experienced hallucinations and confusion, but no long-term ill effects.

 Note People who miss sleep for several days begin to experience *microsleep*, a phenomenon where the brain shuts off for a few seconds at a time. This can cause the microsleeper to lose track of a conversation, a multistep task, or a heavy piece of machinery, sometimes with dangerous consequences. Microsleepers are usually unaware of what's happening—if it happens to you, you'll probably notice little more than a vague feeling of zoning out. The best antidote is a decent sleep—soon.

The effects of staying awake even longer are difficult to pin down. Sleep-deprived rats eventually descend into crazed and violent behavior, lose the ability to regulate their body temperature, and die. And while there are no proven cases where human sleep deprivation has led to death entirely on its own, there are some intriguing question marks. Sufferers of a rare genetic disorder called *fatal familial insomnia* (FFI) lose their ability to sleep in late adulthood. They gradually descend into a world of hallucinations, exhaustion, and dementia, and die in a matter of months.

However, the horror of FFI isn't a slam-dunk case for death by insomnia. FFI is marked by a steady disintegration of the *thalamus* (a region of the brain just above the hypothalamus), and the ultimate cause of death is probably not a simple lack of sleep, but a complete failure of the brain's biological clock and the body's ability to manage its daily rhythms. Another rare disease known as *Morvan's syndrome* can lead to severe sleep disruption without death. Researchers watched one sufferer of Morvan's syndrome last months with apparently no sleep, and with no obvious mood or memory disorders (and no feelings of sleepiness). However, he did have a dream-like state of vivid hallucinations that hit every night and lasted about an hour.

So, the honest answer is that there is no conclusive evidence that staying awake can be fatal. But as you'll see in the next section, scrimping on the snoozing isn't without some obvious effects.

Short-Term Sleep Deprivation

Although the effects of long-term sleep deprivation are difficult to pin down, the effects of its short-term counterpart have been studied in exquisite detail by somewhat sadistic researchers. They include:

- **Attention.** If you're short on sleep you can't concentrate. You're likely to be sidetracked more often, and you'll have particular trouble with long, repetitive tasks and anything that needs solid focus—say, mental math.

- **Reaction time.** Sleep-deprived people slow down. Whether you're driving on the freeway, playing a video game, or trying to win a ping pong tournament, you'll perform more poorly. Miss two or three days of sleep, and you're likely to find that your body is less coordinated and your speech is slurred.

- **Mood.** As you'll learn later in this chapter, there's good reason to believe that sleep has a mood-regulating role. When you're starved for sleep, the brain loses its ability to check rampant emotion, leaving you irritable, quick-tempered, and depressed.

- **Weight gain.** When you sleep, leptin levels fall and ghrelin levels rise. (You learned about leptin and ghrelin, two of the hormones that regulate appetite, on page 39.) If you skimp on sleep for long periods of time, you end up with extra ghrelin, less leptin, and a greater appetite. Lack of sleep also boosts stress hormones like cortisol and can increase insulin resistance, both of which can contribute to weight gain.

Significantly, these problems aren't limited to people who miss an entire night of sleep. They also affect you if you miss smaller amounts of sleep over several days. In fact, studies show that regularly missing one or two hours of sleep a night can quickly add up to the same problems that are typically seen after one or two nights of total sleep deprivation. Even worse, people who are short-changing their sleep know they're tired, but don't realize that their sleep debt is adding up to some serious trouble. Shambling about in an all-too-familiar fog, they have no idea that their performance is at rock bottom levels.

The following chart shows the tolls of inadequate sleep, based on a recent study (see *http://tinyurl.com/2mjk97*).

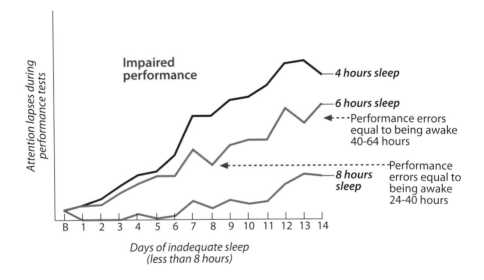

As you well know, the modern world is filled with overworked, sleep-deprived people. The fellow who drives the subway you're riding could well be running on empty. The manager who decides to hire or fire you is probably functioning no better than a person who's on a 72-hour sleep-free bender. Ditto the individual who delivers your mail, manages your finances, defends you in court, cooks your fast-food hamburger, and writes the TV shows you watch. If lack of sleep adds up to poor performance, you can see that we're in a bit of a sticky situation.

The Sleep Cycle

To continue your exploration into the brain and its sleep habits, you need to take a closer look at exactly what your brain does while you're snoozing.

The sleeping brain goes through a cycle that typically lasts about 90 minutes, and repeats that cycle about four times each night. The different stages of the cycle are characterized by dramatically different forms of brain activity. Researchers can spot these stages by hooking a sleeper up to an EEG machine, which records the brain's electrical activity.

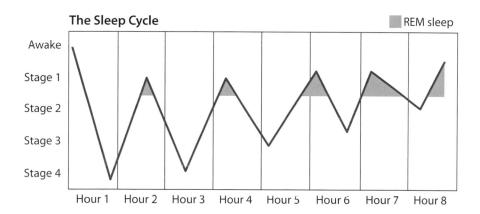

Here's a quick rundown of the sleep stages your brain travels through every night:

- **Stage 1.** This is a drowsy semi-conscious stage. Breathing slows and you may experience *hypnagogic imagery*—visual and auditory hallucinations (for example, flashes of light and sounds of crashing surf) that have no overarching narrative.

- **Stage 2.** This is light sleep. Brain activity slows, but is punctuated by brief spikes of activity called *sleep spindles*, which last one or two seconds. Half of all the hours you spend asleep are spent in this stage.

- **Stage 3.** This is a transitionary period of ever-deepening sleep.

- **Stage 4.** This is the deepest stage of sleep. Heartbeat and blood pressure have slowed, and the brain shows a slow, steady form of activity known as *delta waves*. This is also the stage of sleep when sleepwalking and sleeptalking strike. If you're woken up while in stage 4 sleep, you'll feel groggy and confused.

 Tip The best time to wake up is at the beginning of a sleep cycle, while you're still in stage 1 or stage 2 sleep. If you're getting the recommended 8 hours of sleep, you'll find it easy to wake up between sleep cycles. At this point, sleep is at its lightest, and minor stimulus—a birdsong, a sunrise, a bulging bladder—can nudge you into full wakefulness. By comparison, if you aren't allowing yourself enough time to sleep and you're using an alarm clock to start the day, you may find yourself shocked out of stage 3 or stage 4 sleep. In that case, you're apt to feel like you've fallen under a cement truck.

The most interesting feature of the sleep cycle is what happens when you complete a cycle. At this point, your breathing becomes irregular, and your heart rate and blood pressure rise to levels nearly as high as when you're awake. Your mind begins to churn, catapulting you into the mysterious phenomenon known as *REM sleep*, which you'll explore a little later in this chapter.

The Practical Side of Brain Science

How to Nap

Many experts swear by the brain-boosting power of naps, and recent studies suggest naps can break up the tedium of office work and improve the brain's attention and performance. Naps are generally uncommon in the Western world, where nodding off while the sun's out seems vaguely scandalous. However, new and more stylish napping trends (such as power napping and caffeine naps; more on those in a moment) are gradually making nappers more hip. Recent studies suggest a well placed nap can stave off mental burnout and keep your brain sharp for longer periods of time.

The secret to getting a good nap is to break it off *before* you enter the deepest part of your sleep cycle (stage 3 or 4). If you make the mistake of dipping into the delta waves, you'll wake up groggy and stagger through the rest of the day. Anecdotal evidence suggests that, on average, a twenty minute nap leaves the brain lightly refreshed. Twenty-minute nappers don't fall deeply asleep—instead, they simply dip into a light, trance-like state.

Nappers have different techniques to prevent deep sleep, including brute force (an alarm clock) and innovative thinking (sleeping in an upright position). Some daredevil nappers even down a mug of coffee before settling down to nap, using the caffeine as a sort of alarm clock to wake them up in short order. However, if you're getting a regular good night's sleep, you can probably train yourself to wake up after twenty minutes quite easily. Follow this brief brain pick-me-up with a light snack or cup of tea.

Sleep Through the Ages

Although adults do well on about 8 hours of sleep a day, the sleep requirements at younger ages are greater.

Newborn	Average Hours of Sleep Per Day
Newborn	18
1 month	15–16
3 months	15
6 months	14–15
9 months	14
1 year	13–14
2 years	13
3 years	12
4 years	11 1/2
5 years	11
6 years	11
7 years	10
8 years	10
9 years	9–10
10–17 years	9–11*
Adults	7–8
Elderly	7–8

*Studies also suggest that during adolescence our clock shifts forward, so we're compelled to stay up late and sleep well past when our parents think we should get up.

It's sometimes argued that sleep requirements continue to decline into our elderly years. However, the latest sleep research suggests that sleep requirements for most individuals remain relatively constant through adulthood and old age. However, older people experience more fragmented sleep, with lighter, shorter nighttime rest and more daytime dozing. In other words, if you're a grandpa you're at greater risk of nodding off before the last page of this chapter.

REM Sleep

REM sleep is named after the *rapid eye movements* that sleepers experience. Unlike other stages of sleep, REM sleep is easy to identify. When you're experiencing REM sleep, your eyes dart back and forth under your eyelids. However, the rest of your body is essentially paralyzed, which acts as a safeguard to prevent you from acting out particularly violent dreams.

REM sleep is closely identified with the phenomena of dreaming. If you wake someone up from REM sleep, you're certain to find them experiencing a vivid dream. However, other sleep states also produce dreams. Usually, these are fuzzier, more sedate dreams, and often they're little more than general feelings and soft-focus visions. But occasionally, vivid dreams are reported in non-REM sleep, most commonly at the end of a long sleep indulgence (say, a Sunday morning).

Current science suggests that our biological drive to rest just might have less to do with the tender ministrations of sleep, and more to do with the freewheeling chaos of dreams. Here are some tantalizing reasons to think REM sleep is a critical part of every brain's night:

- When deprived of REM sleep (for example, by being repeatedly woken up in the middle of a sleep cycle), the brain fights back, plunging itself into REM sleep more quickly.

- If you don't get your normal amount of REM sleep in a night, you're brain alters its sleep cycle the next night, spending more time in REM sleep to compensate.

- Adults spend about 20 percent of their sleeping hours in REM sleep. Newborns spend about 50 percent of their time in REM sleep, and fetuses are thought to stay in a nearly-perpetual state of REM sleep. In later years, REM sleep declines to a more modest 15 percent of sleep time. This correlation between periods of heavy brain development and long intervals of REM sleep hints that REM sleep might be playing an important role we haven't quite pinned down.

- Virtually all mammals experience REM sleep. However, REM sleep is a risky time, because it renders animals paralyzed and helpless. To make matters worse, during REM sleep the body consumes nearly as much energy as when it's awake, which is a marked contrast to the other, thriftier stages of sleep. This suggests that REM has some fundamental importance, or a superior race of non-dreamers would have evolved millions of years ago.

However, the picture isn't all neat-and-tidy. Here are some counter-arguments that question the value of REM sleep:

- Studies have deprived people of REM sleep for two or more weeks at a time, with none of the obvious impairments suffered by total sleep deprivation or chronic undersleeping.

- Scientists have the example of an Israeli man who lost the ability to experience REM sleep when a piece of shrapnel was lodged in his brain. Despite this change, he successfully completed law school. (And while becoming a lawyer is by no means an indication of a healthy, functioning brain, no life-changing symptoms were noticed as a result of the lack of REM sleep.)

- Most antidepressant drugs have the side effect of suppressing REM sleep. Again, no ill effects are noticed. However some argue that the suppression of REM sleep may be part of the way these poorly under-stood drugs work.

In the following sections, you'll take a brief trip to the cutting-edge world of dream research, and you'll learn about two hot neurological theories that attempt to explain the real reasons for REM sleep and are quickly piling up some impressive supporting evidence.

 Note Although most sleep research focuses on REM sleep, the oddest stage of sleep, it's quite likely that all stages play different, but complementary roles. For example, sleep spindles (one-second bursts of activity seen in stage 1 sleep) may be another way the brain fine-tunes its wiring, rebalances its levels of neurotransmitters, or prepares to encode long-term memories. Similarly, deep stage 4 sleep boosts the immune system, prepares developing bodies for growth, and is essential for healing the body after an injury.

How Sleep Boosts Learning

Neuroscientists have collected some persuasive evidence to suggest that REM sleep helps the brain rewire itself, integrate new memories, and train itself for important tasks. The first hints at this idea came from a few com-pelling (and somewhat gruesome) experiments:

- In 1959, a French neurologist named Michel Jouvet cut the neurons that cause paralysis during REM sleep in the brains of a few unlucky cats. Then he watched them in their sleep as they got up, cleaned themselves, stalked imaginary mice, fought imaginary enemies, and rehearsed attacks and retreats. He concluded that this process was a kind of practice that would polish these behaviors, possibly making the difference between life and death (or catching a meal and going hungry) in real, waking life.

- In 2001, another study forced rats to run through a maze (much like thousands of similar studies before it). However, these rats were different—they had electrodes implanted in their brains that recorded the activity of some of the neurons in their brains. When these rats took a break from maze running and fell into REM sleep, their brains replayed the exact same pattern of electrical activity as when they were running the maze. This discovery suggests that REM sleep isn't simply a time to practice instinctive behaviors, but a time to hone newly acquired skills.

 Note As you know from Chapter 1, you are what you do—in other words, the more often a group of neurons fires, the stronger and more tightly wired they become. Thus, by replaying their maze experience, rats might firm up their maze running prowess.

- A third experiment bolstered the case by putting humans under the microscope. This experiment didn't involve cutting or implanting anything in the brain. Instead, the experimenters asked their subjects to play the block-positioning video game Tetris. (Clearly, there are some times when it's better to be a human than a cat or rat.) They then monitored the sleep of their subjects, waking them up to confirm that Tetris made an appearance. Even the few amnesiacs in the study dreamt of columns of drifting blocks, although they were at a loss to explain what the images meant.

Additional studies have found that those who get the most REM sleep improve the most in tasks like Tetris playing. People who are taught a new task and then prevented from sinking into REM sleep (in other words, interrupted as each sleep cycle draws to a close), fall behind. However, the strength of this effect is hotly debated.

There's also a catch. REM sleep appears to boost performance with tasks involving *procedural memory*—the subconscious *how-to* knowledge that you call upon when walking, riding a bike, and performing most physical tasks. This may be part of the reason that musicians practicing a new piece struggle through practice during the daytime, and suddenly find they've mastered it a few days later with no extra work.

 Tip If you're learning a new sport or are practicing a new piece of music, make sure you get a solid sleep the following night. This ensures that you'll get a full complement of REM sleep, and gives you the best chance of mastering your new skill subconsciously.

How Sleep Manages Emotions

Some scientists also believe REM sleep is a powerful force for stabilizing moods. There are a few clues that this might be more than just fanciful speculation.

- First, many dream surveys have found that negative emotions rule the show at night, and that fear, anxiety, and guilt crop up relentlessly.

- Second, people who suffer from depression have a dramatically different dreaming pattern in REM sleep. They begin with dreams that seem less emotional, and gradually progress to longer, more negative dreams. They spend more time in REM sleep, and wake up exhausted. It's difficult to determine if this is simply a symptom of depression, or something that actively reinforces the cycle of gloom. However, the fact that antidepressants suppress REM sleep seems to suggest that something may be going wrong in depressed minds when they dream.

If sleep manages emotions, it does so by allowing the emotional detritus of life to be addressed, reconciled, and neutralized. This pattern can also explain the nightmares of *post-traumatic stress disorder*. In such cases the brain may be trying to work through highly-charged negative emotions, but it's consistently interrupted and unable to complete the process of emotional reconciliation.

Although the possible connection between REM sleep and emotions is interesting, it doesn't lend itself to do-it-yourself practical applications. However, here are two good pieces of advice:

- If you're troubled by recurring negative dreams, try rehearsing a positive ending. That new ending can then be incorporated into your dreaming experience, forming new neural connections, and allowing the negative emotions to be dealt with and integrated. (This process appears to happen naturally for many dreamers.)

- If you've suffered from depression before, don't skimp on sleep. As reported earlier, the sleep-deprived brain is an emotional brain. Anger and anxiety run amok, and the pattern of brain activity is surprisingly similar to what's found in people suffering from depression or post-traumatic stress disorder.

Dream Analysis

The most obvious hallmark of REM sleep is irrational dreaming with vivid, hallucinatory detail. However, it's quite possible that REM sleep isn't designed to create dreams. Instead, dreams might simply be a side effect of the lower-level brain processes that are going on during REM sleep (possibly memory consolidation and emotional regulation). As your brain is flooded with a chaotic series of images and memories, the reasoning centers of your brain do what they're trained to do—they struggle to make sense of the disorganized mass of information by weaving it into a barely logical story.

Even if dreams aren't anything more than noise in the higher regions of your brain while the older, more primitive levels do their housekeeping, they can still be mind-bendingly fascinating. Dreams can also be useful, by providing insight into your emotions or giving you a burst of creative thinking.

The Content of Dreams

Although you undoubtedly remember a few of your most memorable dreams, you probably don't have as good an idea about the overall pattern of your dreaming, and how your dreams compare to the nightly visions of other people. Large dream studies shed some light into these questions by comparing the dream journals of hundreds of volunteers, sometimes over long periods of time. Here are some of their discoveries:

- **There's not much sex.** Sure, it happens, but not nearly as often as a lusty Freudian psychiatrist might have you believe. (That said, sex plays a commanding role in daytime fantasies.)

- **Dreams incorporate the ordinary.** Most dreamworld objects, people, and themes come straight out of personal experience. Often, it's recent experience, and it's not necessarily important, unusual, or emotionally charged. (For example, I once had a dream about putting away my socks after a day where I did, indeed, put my socks away. Neuroscientists would conclude that this dreary yawn of a dream isn't an indication of a developmentally delayed brain, but a perfectly sensible example of ordinary dreaming.)

- **Average dreams are forgotten.** The dreams we remember are the dreams that are most memorable. These dreams aren't a good sample of your full nightly repertoire of dreams. For example, alarming dreams that wake you up are most likely to be remembered. Dreams that are exciting enough to tell your friends (often involving flying, fighting, and romantic conquests) also jump to the top of the list.

- **Dreams feature common themes.** Dreams may seem to be the ultimate example of personal self-expression, but many of their themes are surprisingly universal. For example, people in hospitals dream about losing control over their lives and dying, students dream about showing up to an exam with an empty head and no underwear, and so on.

- **Over the course of your life, dream themes recur.** In fact, you probably already have a personal hit parade of dream themes that you might not even be aware of. For example, some people are continually looking for things in their dreams, others are trying to get more attention, some are always engaged in family dramas, while others are perpetually on the run.

- **Toddler dreams are dull.** Despite experiencing more REM sleep, very young children appear to have less vivid dreams. It's possible that the higher reasoning centers that give dreams their strange layers of plot, meaning, and association aren't sufficiently developed in our early years, or they aren't completely wired up yet.

Although it's fun to tease intricate meanings out of the most surreal and convoluted dreams, most neuroscientists will tell you not to bother. Always eager to throw cold water on a little fun, they'll tell you that studying a dream is like analyzing the results of a free-wheeling brainstorming session.

 Tip By examining the details in your dreams you can learn something about the brain that created them. But unless your life holds some deeply repressed trauma, there's no reason to assume your brain is couching its true feelings in elaborate symbolism.

Keeping a Dream Journal

Long before television existed, the human race had ample access to sex, violence, and impossible-to-accept plot twists in the world of dreams. It's no surprise that some enterprising minds want to harness this wellspring of creativity, so they can draw upon its inspiration in the cold, dreary light of a day-do-day reality.

One of the best ways to collect the insight in your dreams is with a dream journal. Here are some tips to make it work:

- Stash your dream journal next to your bed (or under your pillow) and perfect the art of writing in the dark while half asleep.

- Be wary of asking too many questions as you attempt to reconstruct a dream—it's all too easy to crush a wisp of dream out of existence by attempting to impose the logical framework of the waking brain.

- Date your dream journal entries so you can see how your dreaming patterns change and what themes recur.

- If you want to collect a lot of dreams in a short amount of time, time your sleep cycle and use an alarm clock to wake yourself out of REM sleep. (Or, just set it to disturb your sleep every hour.) This technique can help you fill an entire dream journal in a night—but don't expect to get a decent sleep.

Late Night Deep Thoughts

Sleep As an Alternate Reality

Most of us assume that our conscious selves go on hold at the end of the day, and pick up 8 hours later. But what if dreaming is a phenomenon that's just as visceral and immediate as consciousness is during the day?

Ordinarily, we forget the vast majority of the dreams we have, unless we're jolted awake. But this doesn't mean we experience our dreams any less vividly than we experience real life. It simply suggests that our memory storage systems are offline while asleep. In fact, our experience of dreaming just might be similar to the way a severe amnesiac experiences the real world (see page 99)—real at the time, but quickly forgotten.

Given the fact that you spend a total of roughly 20 years asleep, it's enough to make you wonder exactly what fills up those forgotten years. And can you truly say you've had a life well lived if you can't account for nearly one third of your time on Earth?

---

4 Perception

Your brain is a reality-construction machine. It takes the vast oceans of information that flood your senses, and transforms them into a highly subjective inner world.

This inner world has a few things in common with outside reality, but less than you'd think. It's run by a processing system that's quick to jump to conclusions, confidently ignorant of its mistakes, and easily fooled. This processing system sees what it expects to see, hears what it expects to hear, and petulantly refuses to be corrected on even the simplest point. You may enjoy this world or you may not. However, you'll never get a chance to step out of your head and take a clear look at what's *really* happening outside.

That's where this chapter fits in. Here, you'll explore some of the ways that the brain shapes outside reality. You'll learn about the quirks of the eyes, ears, and other senses, and the automatic assumptions that are deeply ingrained in your brain. Occasionally, this knowledge will help you "unfool" yourself—in other words, it lets you anticipate your brain's hiccups and work around them. Other times you'll learn enough to fool someone else, which is just as good (and makes a solid foundation for a career in politics, advertising, or real estate). Either way, this chapter gives you an opportunity to pull back the curtain and steal another quick look at the strange machine that runs your life.

The Doors of Perception

It's tempting to divide the brain's information processing into two neat categories: conscious (what you *know* you see and hear) and subconscious (what your brain deals with automatically, behind the scenes). After all, you don't consciously perceive the inner ear signals that ensure you stay balanced while navigating an intricate dance routine, but you are acutely aware of the crushing heel that your dance partner just placed on your big toe.

However, if you dig a little deeper into the brain's jelly-like matter you'll quickly find that it's a little bit like sharing an apartment with a group of freewheeling friends—there's a lot more going on than you realize (and a fair bit more than you'd probably consent to). Basic avenues of perception that you take for granted, like seeing, hearing, and touch, are actually colored by layers and layers of the brain's automatic preprocessing. In essence, your brain expects the world to behave in certain ways, and it subtly shapes your perception according to these biases.

Furthermore, this isn't just a story about any one sense. It most obviously affects vision, but its effects are equally apparent with sound, touch, taste, and more complex combinations. These automatic assumptions happen at the lower levels of the brain (for example, through specialized neurons that deal with particular optical phenomena) and higher ones (for example, in the folds of the cerebral cortex, where deep thinking takes place).

Although this automatic processing sounds a bit suspicious, you'd be ill advised to turn it off (and short of heavy quantities of illegal pharmaceuticals, there's no way you could). Most people don't want to spend minutes thinking about shapes, illuminations, and perspective simply to follow their favorite sitcom. Similarly, they don't want to go through a painstaking process of logical deduction to determine if the object they're looking at is a person and, furthermore, if it is in fact their spouse (as memorably described in Oliver Sacks' *The Man Who Mistook his Wife for a Hat* [Summit Books, 1985]).

That's not to say it isn't worthwhile to learn more about the automatic processing of your brain. Using the insight you pick up in this chapter, you'll be able to:

- **Defend yourself against accidental mistakes.** A little bit of knowledge can help make certain that you aren't tripped up by faulty brain assumptions (or at least figure out what went wrong after the fact). This is a theme you'll revisit throughout this book, including the next chapter, when you'll discover the ways the brain can mangle memories despite the best intentions of the rememberer.

- **Defend yourself against out-and-out trickery.** Magicians, pickpockets, and psychics often rely on the well known quirks of human perception—the assumptions, omissions, and unusual glitches the brain encounters while processing the outside world. Once you know what to expect, you'll be able to unravel a few tricks (or get better at pulling them off yourself).

- **Dazzle your friends with party tricks.** What über-geek doesn't need a trusty optical illusion to break the ice at a party? And if your interests are more practical, wagering possibilities abound ("Are you willing to bet this line is longer than that one?").

Late Night Deep Thoughts

Can We Know Reality?

We often feel that our senses "wire" us into the outside world. They can be a bit dodgy at times, but surely they're based on objective reality—that is, the sounds and sights that are actually unfolding around us.

This comfortable assumption doesn't account for the many ways that human biology shapes the world we perceive. The most obvious example is the way we perceive colors. Due to the architecture of the human eye (which contains three types of color-sensing cells), the continuous spectrum of light is split into arbitrary regions that have dramatically different meanings to painters, drag queens, and interior decorators. The rather embarrassing truth is that there's no qualitative difference between red wavelengths of light and blue wavelengths of light—they're simply different parts of a smooth, continuous spectrum of light. You could make just as strong a case for breaking the spectrum into 12 primary colors as three. It's somewhat like calling some water green and other water yellow based on its temperature.

In fact, there's no reason why human beings couldn't have crossed the millennia of evolution with dramatically different optical hardware—for example, eyes that divide light into two dozen colors or eyes that can tune into parts of the light spectrum we normally can't discern. All this suggests that our senses give us a projection of the external world—one that filters most of it out and shapes the rest in more meaningful (to us) human terms.

And color is only the beginning of the trouble. A similar problem exists with our hearing, which translates something purely mechanical—vibrations of air molecules—into speech, music, and irritating late-night car alarms. The next time you're listing to a dull sermon from your boss, remind yourself that sound is essentially a "made-up" phenomenon. There's no reason that human beings couldn't sport some other type of organ to transform a limited band of input (say, underarm tickles) into a completely different way of perceiving the world. After all, our world is filled with stimuli we can't perceive—infrared (heat) radiation, ultraviolet light, electromagnetic fields, you name it. For the most part, we aren't tuned into these details because they don't give a competitive advantage to life on Earth. However, they're no less fundamental a reflection of reality than the narrow sliver we do perceive.

Optical Illusions

One of the most fascinating ways to size up the workings of the brain is by exploring *optical illusions*, the strange images that aren't quite what they seem to be. To a certain extent, all optical illusions work by exploiting a chink in the brain's visual processing systems—an automatic assumption that doesn't always hold true, an interpretive technique that can run astray, an attempt to compensate for another shortcoming, and so on. However, there's an amazing diversity in the way these illusions work. You can easily line up a dozen different optical illusions and find that each one relies on a different trick to short-circuit the brain.

Some of the simplest illusions work by *overstimulating* some part of the brain's visual processing system. Conceptually, their effects are like the afterimage you get when you stare foolishly into the sun (against your mother's advice).

One example of this phenomenon is found in the grid of squares shown below. When you stare at it, you'll see gray shaded areas flash into existence where the white lines intersect, even though there's nothing there.

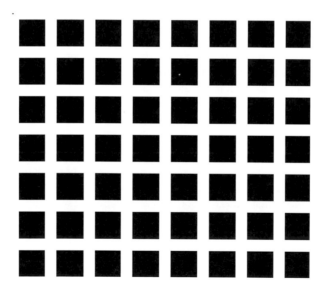

As with many optical illusions, it's difficult to pinpoint exactly what goes wrong in your brain when you look at the grid. However, part of the brain's strategy when picking out shapes involves emphasizing edges and contrasts. In high-contrast images like this grid and the slanted lines shown below, the effect can be pumped up to dizzying proportions.

In order to perceive a scene, your brain takes the information from your eyes and pushes it through a long, complex pipeline. (Actually, the *pipeline* metaphor isn't quite correct, because it implies that operations take place sequentially. In reality, your brain has many visual modules working at the same time, sometimes collaborating to arrive at an insight, other times competing to decide the best interpretation of what you see.) The illusions shown here kick in at a low level, before your brain has a chance to process the full details of the scene in front of you. Although they make for fun eye candy, they don't teach us very much. They're also short on practical payoff, unless you're planning to disorient friends and colleagues with bursts of random patterns.

In this chapter, you'll see a variety of optical illusions, including some that game the system early on (like the grid illusion) and others that mislead neurons further down. You'll also consider illusions that cross over into your other senses, such as hearing and touch. All of these illusions emphasize the same sorry fact—namely, your brain is a very unreliable spectator.

Note One of the oldest known illusions is based on touch, and was described by Aristotle more than 2,000 years ago. Happily, you can pull this one off at home. First, find a pencil, and lay it in front of you. Then, cross your middle and index fingers (they're right next to each other, so that's easy). Now, without looking at the pencil, lay both your fingers on it. You'll have the distinctly odd impression that there are two pencils. (That's why you avert your eyes. If you look at the pencil, you give your brain a chance to correct itself.)

Your Shifty Eyes

Some of the most captivating optical illusions are those that involve imaginary motion. Like the pattern of dots shown to the right, they appear to undulate hypnotically.

This illusion packs in two tricks. First, it uses contrasting colors that are perceived by different cells in the eye (without which the effect is much more subdued). Second, it varies the shading of different dots, placing the shadows above, below, and to the side of the various dots. (This trick is duplicated in hundreds of optical illusions.) However, neither of these details explains how a static image can fool your brain into seeing nauseating motion.

To really understand this illusion, you need to realize that your eye has a dirty secret—it's only able to see fine detail in a small fragment of its visual field. The pinpoint-sized part of your eye that sees sharply is called the *fovea*. If you look at a person an arm's length away, the fovea gives you a sharply detailed region that's about the size of a dime.

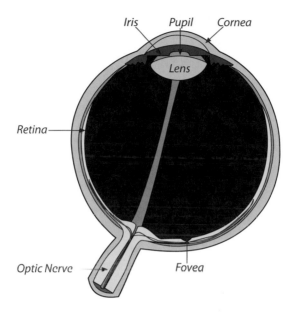

Your brain uses a crafty trick called *saccades* to compensate for this weakness. Saccades are quick, automatic eye movements. They're keenly important for reading books like this one, and they're equally indispensable for taking in the full detail of a visual scene. On average, your eye performs two or three saccades each second, ricocheting about your visual field without you even realizing it, each time capturing the fine detail of another tiny region. Inside your brain, these separate dime-sized pictures are pasted together to create a single, seamless whole.

 If you're severely drunk, your saccades slow down, and you start to see the world the way your eye really perceives it—a patch of sharpness surrounded by a blurry field.

With this in mind, the drifting dots you saw earlier are easier to understand. As your eyes jump from one circle to the next, trying to stitch together the complete picture, your brain is confused by the alternate shading. After each saccade, the previously viewed dots aren't quite where your brain expects them to be, and so it assumes that they've shifted ever-so-slightly to the side. This creates the impression of motion.

You can will away this illusion to a certain extent by focusing intently on a small section of the pattern, and refusing to move your eyes. In this case, the center stops moving, but the sides continue to swell and heave like an unsettled sea.

 Note Saccades don't just compensate for the blurriness of your vision outside the fovea. They also compensate for the unequal distribution of color sensitivity in your eye, and they mask your blind spot (which exists where the bundle of optic nerves exits your eye on the way to the brain).

One of the most famous illusions to take advantage of our shifty-eyed nature is the rotating snakes illusion, created by Akiyoshi Kitaoka, and shown in many alternate incarnations at *www.psy.ritsumei.ac.jp/~akitaoka/rotsnakee.html*. The effect is strongest out of your peripheral vision.

Saccades are one of the many ways your eyes can deceive you. During a saccade, your brain compensates for the sudden movement by temporarily shutting down your visual input. This ensures that you don't see a dizzying blur streaking across your field of view. However, it also means you can miss sudden events that happen during a saccade (just as you can miss something important when you blink). Not only do the eyes lie, they also omit.

Magicians, Psychics, Thieves, and Advertisers

They all have one thing in common. They realize that your brain chooses what you look at. Your conscious mind may influence that decision, but it's all too quick to cede control to the more instinctive, automatic centers of your brain.

Magicians and pickpockets know that sudden movement, sudden light, and sudden noise will always divert your attention, even if you're trying to concentrate. To understand why, it helps to understand that the visual input that's funneled in through your eyes is sent to different regions of the brain. While the higher processing regions in your *cerebral cortex* are busy puzzling out the scene, a more primitive region called the *superior colliculus*, which sits at the ancient core of the brain, scans for signs of danger.

The superior colliculus isn't interested in interpreting what it's seeing in any detail—in fact, it doesn't get the detailed visual information that's passed to other parts of your brain. Instead, it simply reacts to a potentially threatening stimulus. If it sees sudden movement or sound, it will direct your attention to the new stimulus, so you can assess if it poses a danger. If it's startled by something more dramatic—a popped balloon next to you or the sudden appearance of a person looming over your shoulder—it may trigger you to jump up, shout out, or start running. Fascinatingly, it's possible to damage the parts of the brain that perform higher-level visual processing while leaving the superior colliculus intact, which causes a phenomenon called *blindsight*. In this state, people believe they are blind, but can still avoid obstacles and react to movement using the subconscious parts of their brains.

The superior colliculus explains how magicians, psychics, and pickpockets can direct your attention away from a nefarious bit of sleight-of-hand. But what about advertisers? Rather than attempt to divert your focus, they use the same techniques to grab your unwilling attention, drawing your eyes to a garish animated billboard or a television commercial that's several decibels louder than the show that preceded it.

Keeping Focused

As you've seen, your brain keeps your gaze on the move, shifting your eyes to take in a full scene and moving your head to fix on important-seeming objects. This automatic movement creates a sticky problem. It means that it's much more difficult to focus on something that doesn't use the dynamics of sound, flashes of light, and bursts of movement to catch and hold your attention (for example, *Gorillas in the Mist* the book, rather than *King Kong* the movie).

This difference is particularly prominent in many business environments, where distractions abound and everything you're expected to do is monumentally boring. In this situation, focusing on a task like data entry is an epic battle between the paranoid parts of your brain, which are constantly on alert and waiting for the cues that indicate danger, and the conscious parts, which just want to get the job done so you can head off to the pub. So what can you do to win the war and keep your attention where you want it?

First, recognize what you can't change. Studies show that it's all but impossible for the brain to tune out distractions by sheer willpower. In other words, if people are given a task and told to ignore something unrelated, they can't. For example, experiments show that if you work on a computer monitor that has a background with a slowly moving starfield, the part of your brain that processes motion remains continuously active. Or, if you're shown pictures of famous faces while working on word problems, the face-recognition region of your brain lights up like a Christmas tree. The same is true when unimportant sound intrudes on your senses—whether it's a ringing telephone or a foul-mouthed coworker, it all gets processed. This is annoyingly inconvenient, but it makes sense. In our deep evolutionary past, tuning out a sound as loud as a hip-hop cellphone ringtone was likely to get you eaten.

With this in mind, here are a few good tips to keep your brain on task:

- **Don't try to fight distractions; eliminate them.** Unplug your phone, turn off your radio, and close the door to your workroom. If you insist on doing your taxes in front of the television, you're asking for an audit. It's a skewed battle because the television has the help of your superior colliculus to reel you in.

- **Make boring tasks just a little bit harder.** Studies show that the brain will start to tune out some superfluous information when it's wrestling with a challenge. (In the previously described studies, that means the parts of the brain that would ordinarily process the starfield's motion or the famous faces become less active when you're struggling with a tough task.) Obviously, this advice only lends itself to certain chores.

For example, if you have to type a long list of names into a computer, you may be able to better keep your focus by racing against a clock, challenging yourself to enter names in larger batches, or playing a risqué rhyming game with each person's middle name.

- **Resist the distractions you can control.** Although the automatic processing of your brain gives us all a certain degree of distractibility, studies suggest that roughly half of the distractions that derail us from tedious tasks are self-generated. Examples include snacking endlessly and hunting down rare action figures on eBay. In the corporate world, some businesses have found that a mandatory email-free day once a week boosts productivity, sometimes dramatically. Another email wrangling option: limit the number of times per day you check (in the morning, at lunch, and an hour before heading out, for example).

- **Don't worry about background noise.** You should be able to tune out continual soft chatter, humming fans, and keyboard typing through a process known as *adaptation*, which is described later in this chapter (page 86). Essentially, the brain adjusts to a continuous stimulus, recognizing that it probably doesn't indicate an immediate threat.

Distortions and Mismeasurements

Many of the most familiar optical illusions are *distortions*. They take advantage of the brain's assumptions to skew the way you perceive contours, lengths, colors, and shading.

For example, the long diagonal lines in the following picture (which run from the top-left to bottom-right) are perfectly parallel. However, the pattern of cross marks in the line fools your brain into thinking they lean toward one another.

Here, your brain is confused by angles that aren't quite what it expects. It's as if your brain expects the hatch marks to cross each line at a right angle. You can almost feel your brain mentally twisting the lines to make them fit its expectation.

The following image shows a more ambitious pattern that easily blinds the brain. The image shows a series of concentric circles, but the brain is locked into a different interpretation, and insists on seeing a spiral. (Trace your finger around one of the circles if you don't believe it's concentric.)

The remarkable part of both these illusions isn't that your brain is fooled—after all, its mistaken logic is reasonable and (more importantly) it's blindingly fast. The amazing part is that even if you carefully measure the angle of the slanted lines or trace out the circles, thereby proving the illusion, you still can't convince your brain that it's made a mistake. In fact, no amount of pleading can convince your brain to alter its wonky interpretation. Your brain may take a lot of rules into account when it decides how to view a scene, but it has no interest in your slow-thinking deductive logic.

 To put it another way, you aren't in control of what you perceive. So expect flaws in your vision and be prepared to be fooled by magicians, UFO sightings, and apparent paranormal phenomena. Seeing may be believing, but only if you don't mind being royally snookered.

Faulty Comparisons

Along with distortions of shape, your brain can also mislead you when sizing up the length, size, and color of an object. And when the brain's assumptions fail, the effects tell us quite a bit about the brain's book of visual rules, tricks, and shortcuts.

For example, the following illusion shows two curved shapes. The bottom shape appears to be larger, but it's actually identical to the top shape.

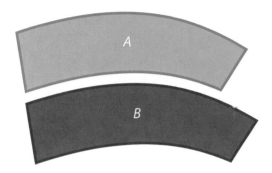

This illusion works because in its haste, your brain makes a few simplifications. It notices the way the left edges of both shapes line up, and takes that into account, discounting the fact that the lined-up edge is gently slanted. When your brain then turns its gaze to the right side, it correctly notices that the bottom shape kicks out a bit further. Thus, the brain concludes that the bottom shape is bigger, missing the fact that the left edge of the bottom shape actually sits a bit further to the right than the edge of the top shape. (If the shapes were truly lined up, the top-left corner of each shape would be positioned on the same vertical line.)

A similar faulty rule is on display in the orange circles of the next optical illusion.

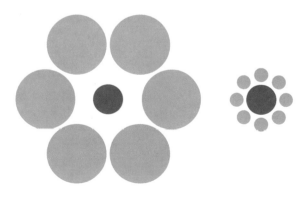

Here, your brain makes two correct observations: the orange circle on the left is small compared to its blue neighbors, and its counterpart on the right is large compared to its neighbors. However, once the brain settles on this intriguing fact, it becomes blind to the fact that both orange circles are the same size. Instead, the proportionally larger one (on the right) seems larger than the one on the left.

The Practical Side of Brain Science

Deceptive Packages

The distortion and faulty comparison illusions are among some of the most profitable for business. These illusions lie behind the tapered packages of shampoo and ice cream that are constantly being reworked to look just as large while holding ever less.

Here are a few of many products with packaging that can betray you:

- Condiment bottles with ridiculously long necks. The brain is better at judging size (the area a shape takes on your eyeball) than volume (the space a container actually has to hold ketchup).
- Bottles of maple syrup that bulge out pleasantly in the middle (where you're most likely to look) but narrow dramatically at the base.
- Sticks of antiperspirant that tower impressively high, while being whittled down to a thickness of a few microns.
- Packages with multiples of anything. Often these packages use carefully designed windows to show you some of the items inside, and use artful contours to imply there are more items inside. When you open the package, you find less than you expect. Your brain's expectation is based on what will fit in the package, but the package designer is more interested in maximizing profit than efficiency.
- Gift baskets supported with vast quantities of unseen tissue paper. Again, your brain sizes up the overall shape and size of a product when judging whether it's worth a second or third look.

When shopping, don't rely on your vision to make a final purchasing decision. Fortunately, most products are required to have key facts stamped on their packages (like weight). Although studying this information won't lead your brain to see the package any differently, it's not your visual centers that pick up the super sleek styling gel and shell out for it at the cash register.

Color Confusion

Shapes and sizes aren't the only thing that can confuse your brain. Your brain can easily make similar mistakes when comparing brightness or colors. In the carefully set up illusion below, two cubes are shown with different lighting. In the center of the front face of the cube is a square that appears to be yellow in one figure and blue in the other. However, the color is actually identical in both—it's the dull shade of gray that's shown in the bar below. (Virtually the only way to convince yourself of this illusion is to use bits of paper to block out the rest of the picture, so that you see only the squares in question.)

In this illusion, the brain isn't exactly wrong—it's simply compensating for what it believes is a difference in lighting. It concludes that a square that appears gray under blue light is probably yellow, and a square that appears gray under a yellow light is probably blue. In other words, your brain's perception has a built-in routine for evaluating lighting conditions. This is the reason you can see quite normally at home in the evening, even though the artificial lights you're using cast a yellow-red shade of light that's dramatically different than the blue-tinted radiance of the sun at noon.

Cameras can't adjust themselves to compensate for the color of light. This is one of the reasons why it's much easier for your eyes to interpret a scene than for your camera to take a great picture. Your brain can smooth out the oddities and inconsistencies of lighting conditions, but film (or the electronic sensor in your digital camera) isn't as forgiving.

A similar effect is at work in the legendary same-color illusion. Here, two squares that are filled with *exactly* the same shade of gray (A and B) appear to be dramatically different. Once again, it's almost impossible to accept this illusion unless you cover up almost everything else in the picture except the two squares in question.

The remarkable part of this illusion is that the brain picks up on a range of clues to make an emphatic conclusion—everything from the 3-D shape of the cylinder that casts the shadow to the pattern of the checkerboard, which darkens significantly but imperceptibly around square B. (The latter part is the most significant factor in the illusion. The brain is deeply attached to the idea of a regular checkerboard pattern, and prefers to see that over anything else.)

The 3-D World

So far, you've seen how the brain has built-in assumptions that help it interpret shapes, sizes, and colors (and sometimes lead to quirky mistakes). The brain also has a bag of tricks that it uses to convert the 2-D image that's projected on your eye to a realistic understanding of the 3-D world in front of you.

Consider the classic example of two lines, shown below. Even though a ruler will tell you that the lines are the same length, the brain stubbornly insists that the top one is shorter.

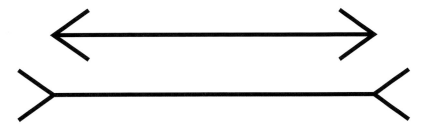

One explanation for this illusion is that the brain is biased towards picking out the cues of 3-D objects. Lines that angle inward are typically seen in objects that are nearby (like the table in the picture below). Lines that angle outward are more common in distant objects (like the back corners of the room). Here's an example that illustrates by comparing two lines that have the same length, but are placed in two different spots in a 3-D scene.

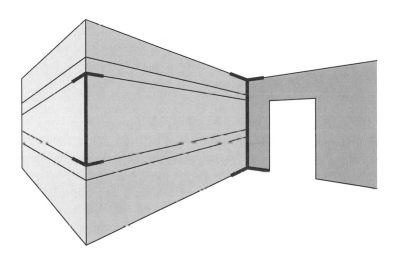

In the two-lines illusion, your brain is well aware of the fact that both lines are really and truthfully the same length. However, your brain also believes that the bottom-most line is farther away. If two objects look the same in your eye, but one is farther away, there's only one possible conclusion—namely, the object that's farther away is bigger. Thus, the brain "corrects" the length of the second line to take the imagined distance into account.

At first, it seems odd that the brain is so willing to skew the size of things based on their perceived distance. However, on second thought it makes a lot of sense. If the brain didn't perform this automatic adjustment, your father would appear to shrink to midget size as soon as he began walking away from you.

The brain has several other tricks for translating the 2-D picture in your eye into a 3-D model. It assumes that objects close to the horizon are farther away, and it compares unknown objects against nearby known objects to infer distance.

 Note A similar effect underpins the horizon moon illusion. In this scenario, the moon appears to be much larger than usual when it's low in the sky. This is because the brain sees the moon in relation to distant objects and the horizon. But when the moon is high in the sky, the brain has no such frame of reference, and so the moon appears tiny and insignificant.

Another 3-D cue is shading. When the brain takes in a scene, it expects to find a sun-like light source radiating from above, and it uses patterns of shading to infer contours and shapes. Humans co-opt these automatic assumptions with artful applications of makeup. To an unbiased observer (say, a computer or an alien being from another planet), makeup would seem like little more than face paint. But for the easily influenced human brain, makeup is processed like shadows, and suggests a more sharply defined face.

Lastly, the brain uses one physical detail to see in three dimensions: the slightly different vantage point that's provided by each of your two eyes. You probably already knew this, but it's a less important factor than you probably thought. The separation of your eyes helps your brain accurately judge depth for very close objects, but it's useless for far off ones. As you can readily test, if you cover one eye and wander around the house you might have trouble doing some precision tasks (like tying a knot or chopping tomatoes), but you won't have any difficulty interpreting the shapes around you as 3-D objects.

Seeing Things

One of the hardest challenges for the brain's visual systems is picking out shapes. It's an extraordinarily difficult task. Shapes can not only be moved, rotated, resized, distorted, and obscured, but they can also exist in an endless number of variations.

The brain deals with this problem using a toolkit of assumptions. And the brain does a good job—it can easily beat computerized shape-spotters when scanning pictures, faces, and moving scenes. However, the brain's eagerness to find shapes also leads it to find shapes where there aren't any, as with the white triangle at the forefront of the following picture.

When confronted with this picture, your brain doesn't **need** to conjure up a white triangle. There's a reasonable alternate explanation—that the image contains three pacman-like circles with wedges cut out of them, and the wedges are lined up with the gaps between the blue triangles inside. However, a just-so arrangement like this would be unlikely in the natural world, so your brain quickly dismisses that possibility. In essence, your brain picks up on a few clues and performs a rapid analysis to determine the most likely explanation. However, you don't merely **think** about that most likely explanation, you also **see** it.

If you rotate the pacmen around, the illusion disappears, and the image reverts to a collection of harmless shapes.

The brain's obsessive pattern matching isn't limited to shapes. It happens with faces (which we see in unlikely places like house fronts and :) punctuation) and speech sounds (for example, if parts of a word are beeped out in a recording, we "hear" the full word based on what makes sense in the context of a sentence).

The brain is also primed to identify letters and spot words. Can you read this sentence?

I LIKE IUMRING TO CONCIUSIONS

If you said "I LIKE IUMRING TQ GQNGIUSIQNS," you have a perverse sense of humor. However, you're also entirely correct.

The Practical Side of Brain Science

The Continual Search for Patterns

As the previous illusion shows, the brain invents shapes and identifiable *things* to explain odd patterns and arrangements. Think of it this way—the brain is in many respects a giant pattern-matching machine. When tuning a pattern-matching machine, you can make it more conservative (in which case it will occasionally miss things) or more aggressive (in which case it will occasionally invent them). The brain does both, although it's more likely to imagine something into existence than block it out, because that proves to be a safer survival strategy.

The imaginary white triangle is one example where the brain fills in some logical ingredients to complete the picture. Here are a few more cases where the brain adds something that might not be there:

- Clouds. Surely searching for recognizable shapes in the sky is just a pleasant way to let the brain's shape-spotting system run rampant—and see what it comes up with.

- Jesus grilled-cheese sandwiches. Oft-sold on eBay, food items that appear to have a holy figure seared into them are most likely the product of an over-eager brain. But you already knew that.

- Fashion. The appeal of some fashions and many undergarments is that, by concealing the exact shapes of the wearer's body, they give ample room for the brain to imagine the—ahem—idealized representation it expects. And no, it's not a white triangle.

Making Something out of Nothing

The brain doesn't just organize seemingly unrelated input into patterns; it also has a nasty habit of imagining something into existence. The dubious Rorschach inkblot test is a good example. Take a look at the card shown below:

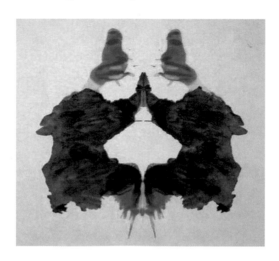

What do you see—a masked face, two bears exchanging a high five, or a meaningless splatter of red and black ink?

When looking at an inkblot, most of us are aware of the imaginative power we're investing to transform ambiguous input into a meaningful picture. However, there are many cases where your brain performs the same operation without you realizing the creative leap that it's taking.

Have you ever thought you heard a telephone ringing or a person calling your name while running a noisy appliance like a vacuum? This effect Is spurred by the brain's pattern-matching systems, which run wild when hunting through a din of sound. A great illustration of this phenomenon is an experiment that asked volunteers to determine when Bing Crosby's *White Christmas* began to play in the background of a noisy recording. The devious trick was that there was no *White Christmas*—only thirty seconds of white noise. But primed with the expectation of hearing the familiar tune, about a third of the participants reported that they heard it. Incidentally, some people seem to be more susceptible to imaginative illusions like this one. It's thought that they have more creative, fantasy-based minds, which are perfect for free-wheeling brainstorming sessions but not as good at skeptical inquiry.

Tip The authors of the book *Mind Hacks* (O'Reilly, 2004) describe the *White Christmas* illusion, and have provided a noisy recording that you can use to test your friends at *www.mindhacks.com/book/48/whitechristmas.mp3*.

The white noise study was very small and far from conclusive. Other studies have argued that people who believe in ESP are more likely to find meaningful patterns in random arrangements of dots. In other words:

[Noisy Input with Little Obvious Information] + [Your Brain] =
[Things That Go Bump in the Night]

The brain's over-eager pattern-matching system may be a plausible explanation for UFO sightings, ghosts, and other late-night creepies.

Ignoring Things

Your brain has another skill that's just as important as finding patterns in the chaos. Not only can your brain imagine new objects into existence, it can also block out the things it wants to ignore.

As you learned earlier, your brain is hard-wired to focus attention on threatening sights and sounds. In order to better separate these potentially dangerous cues, the brain filters out repetitive, unchanging stimuli like a whirring air conditioner or the rocking motion of a boat at sea.

 Note There are many different neurological processes supporting this "tune-out" behavior. At the lowest level, constantly stimulated neurons temporarily stop firing. (For this reason, your eyes jitter imperceptibly back and forth even when you hold your gaze steady. If they didn't, the same neurons would always be stimulated by the sight in front of you. They'd get tired out, stop firing, and everything would fade out into blackness until you looked somewhere else.) The brain also has higher-level processes that adapt to constant stimuli and direct attention away from things that aren't changing in favor of those that are.

Most of the time, your brain's tune-out feature is exactly what you want. After all, who wants to be bothered thinking about the sound of air rushing by your ears, the feeling of weightiness as you sit on your couch, or the tactile sensation of clothes rubbing against your skin? Instead, your brain notices each one of these phenomenon briefly when they first appear, and then quickly adapts to ignore them. However, sometimes this effect can lead to some interesting illusions.

You're no doubt keenly aware of the way the brain adapts itself to different levels of brightness. (If not, try walking from a darkened room into a bright summer day without getting run over.) However, the following version is more fun:

❶ Stand in a doorway, with your arms down at your sides.

❷ Place the back of both hands against the door frame on either side.

❸ Push up with as much strength as you can muster. Keep this up for a couple of minutes.

❹ Now relax and walk away from the door.

For the next few minutes, you'll have the sensation that your arms are drifting up, weightless—in much the same way that your brain might adjust to a stronger gravitational field on another planet. After only a couple of minutes in the doorway, your brain becomes accustomed to the fact that it needs to exert more effort to keep your arms up than at your sides.

As this experiment shows, the brain's tendency to ignore things is really a remarkable ability to adapt itself to its current environment. There are dozens of do-it-to-yourself experiments that show similar adaptations at work. For example, if you scatter your living room furniture haphazardly, you'll spend the first few hours bumping into sofas, the next few hours steering yourself effortlessly (and subconsciously) through the chaos, and the following weeks wondering why everyone looks at you so oddly when you have them over for tea. A similar automatic adjustment and eventual ignoring happens with smells, but even more quickly. If you want to know if the scent from last night's curry cook-off is still around, you'll need to step outside your house and then come back in, because your nose tunes out even the strongest smells after just a few sniffs. And if you want to answer the age-old question "Do I smell OK?" you'll need the help of a friend, because your brain is perpetually filtering out the familiar odor of your own body.

Lastly, your brain also adjusts itself automatically to pleasure, making sure you don't get too much of it no matter how many triple-chocolate sundaes you down in a single sitting. Chapter 6 has more about this frustrating fact.

Other Perception-Distorting Assumptions

You've no doubt seen illusions that use ambiguous pictures, which can be interpreted in different ways. One legendary example is the two faces and vase, shown below.

The interesting thing about this sort of illusion is the fact that when you see it for the first time, you're likely to settle on just one way of seeing it. You'll remain oblivious to the alternative possibility until a smug friend points it out.

This sort of automatic interpretation is obvious with contours and shapes, but it also applies to more complex meanings that we assign at a higher level. In fact, these sorts of snap judgments often color what we see, even though they aren't specifically related to vision.

For example, consider the following figure, which was the subject of a cross-cultural study. If you were asked to express this scene in a couple of sentences, how would you describe it? Think out your answer before continuing.

Most Westerners describe this scene pretty plainly. There's a group of people gathered in discussion (possibly a family), there appears to be a window on the left above one of the women, and the shading of the floor and corner of the wall make it clear that everyone is gathered indoors. But these obvious "facts" aren't quite as obvious to people with different experiences and hence different assumptions engraved in their brains.

When researchers showed this picture to East Africans, nearly all of them said the woman on the left was balancing a box on her head. And the corner of the room in the back was interpreted as a tree, under which the family is sitting. Now if you look at the figure again, you'll probably agree that this interpretation makes just as much sense as your own. As a Westerner who has spent much of your life indoors, your brain is used to interpreting scenes using the boxlike shapes and angular cues of modern architecture (like windows and the corners of walls). Rural East Africans have a different store of experience to bring to bear on new scenes. All this shows that a surprising amount of higher-level reasoning can leak into processes like hearing and seeing, and color the results without you even realizing it.

Dizzy Yourself Silly with Optical Illusions on the Web

You've now finished your tour through the oddities of human perception. But the fun doesn't need to end here.

Once upon a time, the only reliable place for average people to see optical illusions was in a somewhat baroque object known as a book. (If you're reading a non-electronic version of this text, you know what I'm talking about.) Books were wildly popular for many years, before becoming re-placed by electronic pixels. Today, books are mostly remembered by some of the over-16 crowd as an odd early version of the Internet.

No matter what you think about the march of progress and the colonization of Earth by computers, the Internet has been good for optical illusions. There are dozens of Web sites that show optical illusions in all their glory. Here are some of the best:

- **Akiyoshi Illusions.** The legendary creator of the rotating snakes illusion (shown on page 72) also provides pages of hand-crafted illusions. See *www.ritsumei.ac.jp/~akitaoka/index-e.html*.

- **Purves Labs.** This top-notch lab studies optical illusions and offers a see-it-yourself section that includes some of the most remarkable color and brightness illusions ever created. See *www.purveslab.net*.

- **Michael Bach's Illusions.** Many of the illusions on this Web site are outfitted with multimedia extras, such as little movies that move parts of the optical illusion around so you can verify what you really want to know—that two lines are the same length, two shapes are the same color, and so on. See *www.michaelbach.de/ot*.

- **Wikipedia.** The free-for-all online encyclopedia describes a selection of optical illusions, including many of the examples you've seen in this chapter. See *http://en.wikipedia.org/wiki/Optical_illusion*.

- **Mighty Optical Illusions.** Although it's a bit noisy—try to use your brain's ignoring power on the Google ads distributed about the top, side, and bottom—this Web site has a solid illusion catalog, and ranges farther afield to get some interesting examples from real life. See *www.moillusions.com*.

Incidentally, the human race takes advantage of the idiosyncrasies of its visual hardware. Most of us spend hours transforming collections of flickering dots on a screen into an impression of real people. This optical illusion, known as television, is surely one of the most impressive visual ruses ever documented.

The Power of Expectation

Built-in judgments aren't limited to vision. Our assumptions can blind us to facts and lure us into unwitting conjecture in virtually every way that we perceive the outside world. Advertisers and other nefarious ill-doers rely on these errors. Here are some examples:

- **Shocking tastes.** If you've ever put a glass to your mouth expecting to drink milk only to taste orange juice, you've felt the power of expectation. There's a moment of palpable shock when the new sensation defies the brain's expectation, and a brief pause before the brain can perceive what it really is.

- **Sizes and weights.** Many studies show that people confuse size and weight, and let their visual assessment of size influence their impression of weight. One of the simplest ways to see this is by getting several different sized containers (say, a small overnight bag, a suitcase, and a large piece of luggage) and filling them up with materials that weigh the same amount (you can use a pile of heavy books). If you ask someone to judge which is heaviest, they will inevitably pick the smallest one. This is because the smallest one defies the brain expectation that small is light, and thereby makes a bigger impression.

- **Food and fecal matter.** Can you tell the difference? Apparently not by smell alone. People respond differently to an odor that they sniff out of a test tube depending on whether they're told that it's fancy cheese or human waste.

- **Packaging.** In many studies, people's reports of how much they enjoy particular foods (and how much they're willing to spend on them) depend on the packaging and brand name as much as the actual food. In his book *Blink* (Little, Brown, 2005), Malcolm Gladwell describes studies that find people will pay about 10 cents more for ice cream in a pleasingly round container rather than a rectangular one, and will rate the taste of Chef Boyardee products worse if Hector the chef looks like a cartoon character.

All of these examples remind you that perception doesn't lie in the eyes, ears, or any other sense organ—it exists in the brain.

5 Memory

I t doesn't matter whether it's a first kiss or a final exam, all your experiences end up the same way. Once the moment has passed, life's most noteworthy moments get fused into your brain as *memories*. And while the memories may seem sharp and vivid at first, if you poke them twice you'll find that many are as soft as a half-baked bagel.

Few of us take the time to kick back and explore our memories. If you did, you'd probably find a smattering of vivid images submerged in a dense and endless fog. Think of the formative periods of your life. Whether it's the early days of a new job, the first few weeks of parenthood, or a month away from home, odds are you'll have a much easier time describing the general "feeling" of the time than producing a detailed day-by-day account. And what you *do* remember will be subtly yet thoroughly altered to match the assumptions, life outlook, and emotional state of the current you (which may not match the mindset of the person who had the original experience). In other words, memories aren't only fleeting—they're also *alive*, and they degrade, evolve, and adapt over the years you keep them.

The study of memory is one of the central pursuits of neuroscience, and it presents some of the brain's most alluring mysteries. In this chapter, you'll explore how we remember and why we forget. You'll consider the different types of memory, and you'll learn how to make the most of your limited short-term memory storage. You'll meet a man who couldn't remember and one who couldn't forget. Finally, you'll learn a series of handy techniques to help make sure important information sticks to your long-term memory.

The Remembrance of Things Past

Many people assume that memory is a *thing*, somewhat like the spiral of microscopic bumps that stores the music on a CD. But memory is actually the *process* through which your brain is continually transformed by experience. The brain has no hidden tape recorder or secret storage cabinet. Memories aren't faithfully recorded and then retrieved at will—instead, they're incorporated into your brain alongside your ideas, beliefs, temperament, and everything else that makes you *you*.

Before you get any deeper into the specifics of how your memory is stored, you need to realize that there are several types of memory. Although the boundaries aren't always clear cut, it's often useful to separate memory into the following categories:

- **Short-term memory.** Also known as *working memory*, this is the very limited memory store that holds details for a few seconds or a few minutes. The items in this category only stick around as long as you're concentrating on them. Ever wondered, "What was I just thinking about?" Well, your short-term memory just tossed it out the back door.

- **Declarative memory.** Also known as *long-term memory*, this is the permanent and virtually limitless store of facts and events that you accumulate over your lifetime. Although it's convenient to talk about long-term declarative memory as being a single entity, it can be separated into many more specialized types, including memory about specific facts, general concepts, language, and the experiences of your life.

- **Procedural memory.** This is the "how-to" memory that comes into play with physical skills. It lets you drive a car, tie your shoes, and play the ukulele without any conscious effort to recall the information you need. Procedural memory is remarkably durable, nearly unforgettable, and able to survive the ravages of diseases like Alzheimer's. There's not much you can do to improve your ability to form procedural memories, although some studies suggest REM sleep gives your brain a chance to strengthen it (page 59).

None of this explains how things are *forgotten*—in other words, what makes a perfectly good long-term memory unmoor itself and drift off into the void. Current thinking suggests that we forget a whole lot less than we think—instead, we simply lose the ability to retrieve our older, rarely visited memories. It's also possible that the brain uses distinctly different ways to store long-term memories for shorter periods of time (say, a few hours or days). But to conclusively answer the question, scientists need to know more about the neurological processes through which the brain indexes and reassembles memories. Despite dramatic advances, neuroscientists are a long way from this goal.

Short-Term Memory

Short-term memory is the most fleeting type of memory. It holds onto a few chunks of information while you're actively thinking about them. If your attention wanders, the information is tossed out in less than a minute, but you can hold on to details for longer by repeating them continuously. Short-term memory is how you remember the toll-free phone number for a revolutionary piece of exercise equipment between seeing it on a late-night infomercial and reaching the phone.

Short-term memory is notoriously limited. Some believe it holds five to nine chunks of information, others claim it holds just four, but all agree that the total number is a few items short of your grocery list. Scientists also disagree on exactly how short-term memory is stored in the brain. But it is as least partly linked to current electrical activity taking place in your neurons—in other words, the pattern of signal transmission that's ricocheting through your head right now. This is markedly different from long-term memory, which depends on permanent physical changes in your brain.

It's important to realize that the five or so pieces of information held in short-term memory aren't detailed concepts. More accurately, short-term memory holds pointers to the more detailed conceptual information that's stored permanently in your brain.

For example, if you think of the items *cat*, *dog*, and *zucchini*, you don't actually have a full conceptual representation of any of these items in your short-term memory. Instead, you have three *links*. For example, the first item (cat) leads to the neurons that encode your long-term understanding of a small, carnivorous, and highly manipulative life form that's related to the lion, but has discovered a way to coax many more calories out of a single human. You have no hope of holding all this information in your short-term memory at once, but with the basic link in place you're able to draw on it and have meaningful thoughts.

Chunking

To test your short-term memory, try remembering the following sequence of numbers:

Give yourself a few seconds and then try to copy the list onto a sheet of paper.

You're unlikely to get the whole sequence, but you'll probably surpass the lowest estimates of short-term memory (four or five numbers). If you do particularly well, it's because you're using some form of *chunking* to compress a meaningless series of digits into more concise, and possibly more meaningful, information. For example, try remembering these digits:

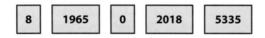

Even though the sequence of numbers is the same, this grouping arrangement is easier to remember. It includes 2 years (1965 and 2018), which reduces eight digits to two chunks. If you try to remember the sequence now, you'll probably have a far easier time. This strategy underlies telephone numbers, which are broken into groups using spaces, parentheses, and dashes to make them chunkier.

This chunking strategy works even better if you can draw on additional information. For example, if 1965 lives in your memory as the year *The Sound of Music* was released, you can store in your head a single chunk that says "*Sound of Music* release date." Then, when you retrieve that chunk from your short-term memory, you'll automatically be able to pull the four numbers (1-9-6-5) out of your long-term memory. A similar trick is to use visual imagery. For example, you can see the final four numbers 5335, as the word SEES written with block digital numbers.

It's possible to remember dozens of digits by compressing them and encoding them as more meaningful information, such as times, dates, words and sentences, and so on. In one test, an otherwise ordinary long-distance runner amazed researchers with his ability to remember up to 73 consecutive digits by chunking them into different run times. However, his short-term memory for other types of information was no better than average, as shown by his very ordinary performance in word-based recall tests.

Later in this chapter, you'll learn some of the best memory techniques for jazzing up ordinary information so it's easier to remember. But first, you need to take a look at how the brain stores long-term memory.

Long-Term Memory

Scientists have spent years searching the brain for the biological equivalent of a memory filling cabinet, where past experiences are stowed away. No such storage place has ever been found. Instead, it now seems certain that the brain remembers by "walking through" massively interconnected groups of concepts that are stored throughout the brain.

For example, say you're trying to remember what you ate for breakfast last week. To dig up the right information, your brain might lead you on a quick tour through your morning routine or a catalog of your favorite foods. Along the way, it asks leading questions (Were you in a hurry? Did you eat alone?) and seamlessly fills in the details you can't remember with some educated guesswork. You're unlikely to notice this process taking place, because the brain can stitch these millions of pieces together into a memory that seems remarkably whole and complete.

How Memories Are Stored

As you learned in Chapter 1, it's relatively rare for new neurons to appear in your brain. However, the structure of your brain is being continuously reworked. *Synapses*—the connections between neurons—are constantly being strengthened or weakened, and new *dendrites* are growing to link neurons together in new patterns. This continuous process of brain reorganization underlies all long-term memory and learning.

 Note A common misconception is assuming that memories are stored in some sort of container in your brain and that your neurons simply pull memories from this storage site when they're needed. A more accurate description goes like this: memory is what's created when specific groups of neurons in your brain fire in specific patterns. In fact, many neuroscientists argue that there's no solid distinction between the act of remembering and the act of thinking.

Although memories are scattered throughout your brain, there is one brain region that plays a key role in coordinating memory storage: the *hippocampus*. The hippocampus is a small neuron bundle that's buried near the bottom of the brain. The human brain has two of them, one on the left side and one on the right.

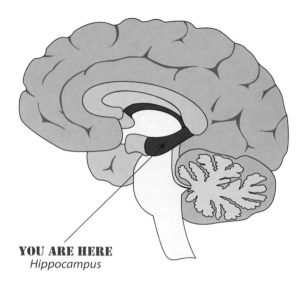

YOU ARE HERE
Hippocampus

The hippocampus has several functions:

- **Navigation and spatial memory.** The hippocampus is keenly important for navigating through mental maps. London taxi drivers, who have to master "The Knowledge"—a grueling understanding of the weaving streets of London, their intersections, and landmarks—have different hippocampi than average folks, with one region that's noticeably larger than in non-cabbies. (The hippocampus is also one of the few areas of the brain where new neurons grow.)

- **Memory retrieval.** The hippocampus recalls relatively recent long-term memories, such as those that are a few weeks, months, or even years old. However, the hippocampus isn't required to retrieve older long-term memories, which have been more thoroughly integrated into the brain.

- **Memory formation.** The hippocampus lets you create new long-term memories. Scientists know this, because they removed it from the brain of a person named Henry M., and his life was forever transformed.

The Man Who Couldn't Remember

Even the most socially awkward neurologist can become the life of the party by telling the story of Henry M., a now-elderly man whose life is forever frozen in 1953—when he had radical *hippocampus*-removing brain surgery. Henry's surgery, performed to rid him of debilitating seizures, also had an outlandish side-effect—it removed his ability to form long-term memories.

When asked how old he is (and one assumes Henry gets this question a lot), Henry invariably responds "about thirty." In Henry's world, Harry Truman is always president, black and white television is a novelty, and furniture made out of chrome and vinyl still seems like a good idea.

Henry's short-term memory is essentially unimpaired. As a result, Henry can keep new facts in his head for a matter of minutes. But when his attention wanders, Henry's brain discards whatever it's just assimilated, leaving him no wiser than when he went into surgery decades ago. (The one exception is procedural memory. When taught new physical tasks, Henry can repeat them later, even though he has no conscious memory of learning the skill. That's because this sort of learning doesn't involve the hippocampus.)

 Note The hippocampus isn't fully mature until about 2 years of age—a likely reason we can't remember early childhood events later in life.

By all accounts, Henry is an intelligent, pleasant, and agreeable fellow. Although he has no idea where he is, Henry often has a feeling that something important is wrong, and occasionally worries that it's his fault. Perhaps after half a century of perpetual confusion, never recognizing where he is, dealing with a stream of nosey scientists, and failing to find any familiar faces, places, or objects, Henry's brain has developed a vague but inexpressible sense of his hopeless situation.

As you've seen, the hippocampus is nestled deep inside the brain. As a result, it's difficult to accidentally damage your hippocampus without destroying other critical parts of your brain. However, Henry's condition isn't entirely unique. Equally compelling is the story of Clive Wearing, a British conductor and early music expert who was at the height of his career when a virus damaged his hippocampus.

Clive's story is an unbearable blend of the heart-warming and the heart-rending. With no ability to store new memories, Clive greets his wife passionately every time she enters the room, even if she left only seconds before. (They were married shortly before he fell ill.) However, he's also tormented by the sense that something is terribly wrong, and haunted by the knowledge that he is utterly unable to comprehend what's happened. Without any long-term memory, Clive has no way to organize the events that are taking place around him into any sort of meaningful narrative.

When not otherwise occupied, Clive continuously experiences the feeling that he's just awakened. It's as if his consciousness is a computer that restarts every time his short-term memory gives out. Clive records these episodes in his journals, which stretch into hundreds of pages and read, chillingly, like this:

7:46 am	~~I wake for the first time.~~
7:47 am	~~This illness has been like death till NOW.~~
	~~All senses work.~~
8:07 am	~~I AM awake.~~
8:31 am	~~Now I am really, completely awake.~~
9:06 am	~~Now I am pefectly, overwhelmingly awake.~~
9:34 am	Now I am superlatively, actually awake.

When adding a new entry, Clive crosses out the earlier ones because he can't remember writing them.

Although Clive is unable to remember anything that has happened since his brain was damaged in his late forties, he seems to have a sense of the time that has passed. Upon his "awakenings," he often insists that he's been dead for 20 years, and describes a terrible absence of sensation and thought over the great void he's just left.

 Note You can learn more about Henry M. from the book *Memory's Ghost* (Simon & Schuster, 1995), which was written by one of the few non-scientists to meet him after his surgery. You can learn more about Clive Wearing in the book *Forever Today* (Doubleday, 2005), which was written by his wife.

Can You Have Personality Without Memory?

The cases of Henry M. and Clive Wearing fascinate scientists and ordinary people alike because they pose questions about the nature of personality. For surely, if you lost the ability to synthesize new memories at this exact instant, your entire personality would also be perpetually frozen in time. With no inner narrative and no possibility of maturing, developing your ideas, your temperament, and your outlook on life, you'd be in a state of perpetual limbo. Neither trapped in the past nor able to participate in the future, it would be like waiting at one of the train station stops along life's journey, but stuck there forever.

The Reconstruction of Remembering

In order to remember something, your brain needs to assemble the memory from a vast network of concepts and details. Here's the problem—when you stick a memory back together, what you get is probably not the original picture. In fact, remembering is an act of creative reimagination. That means a glued-together memory doesn't just have a few holes and out-of-place pieces, it also has some entirely new bits pasted in.

Over the past few decades, psychologists have keenly explored the limits and distortions of memory. The following sections explain some of the things they've found. All these points suggest the same conclusion: never trust an unverified memory, whether it's a witness recalling a crime, a know-it-all friend quoting a research study, or Grandpa Sid describing the idyllic prairie life of his youth.

Leading questions shape memories

This principle seems obvious (after all, asking an eyewitness "Where did the creep strike his wife?" is different than "Where was Bill standing when the alleged incident happened?"). However, this effect operates just as strongly when the word variations are subtle. For example, one study showed volunteers a video of colliding cars and then asked them either "How fast were the cars going when they *smashed* into each other?" or "How fast were the cars going when they *contacted* each other?" The alteration of a single word led respondents to estimate dramatically different speeds.

People incorporate new "information" into old memories.

Memories are never frozen moments in time. Instead, they silently suck in new information. For example, in another traffic accident test, some volunteers were asked a question referring to a yield sign. Even though the yield sign didn't exist in the video, they were more likely to describe it in their later recollections. A similar effect occurs when published accounts of a crime contaminate the memory of witnesses.

 Note Memory studies show that you can't always trust an eyewitness to convict or even identify a suspect. Furthermore, several imperfect witnesses don't add up to one reliable witness—in fact, if they've been exposed to the same information, they may be just as confused but far more convincing. Consider the example of the 2002 Washington sniper attacks, where one person's apparently legitimate memory (seeing a white van before a shooting) became integrated into dozens of other tipsters' memories, leading to mass confusion.

People remember "facts" more readily than sources

Do potatoes cause cancer? (For the record, no.) When you read that tabloid headline in a grocery store checkout line, it may make an impression, but you won't be fooled. But if a few months later someone mentions a new study about potatoes, it just might trigger a recollection of the headline you read earlier. Only the odds are that now you'll remember the content of the story, but not its source. You'll be left scratching your head in puzzlement and wondering if it's time to cut the harmless spud out of your diet.

 Note Advertisers love this effect—it's what makes irritating commercials work. Even if you don't believe ridiculous health claims and comparisons against other brands, you're more likely to choose a brand that's become familiar through an unending advertising barrage, and springs readily to mind when you see its image on store shelves.

Rehearsal turns suggestions into memories

Researchers have had entirely too much fun showing how easy it is to implant false memories through the force of suggestion. Famous studies have used this technique to compel volunteers to remember childhood memories of being lost in a mall, spilling a punch bowl at a wedding, eating pizza with a clown on their birthday, and hugging Bugs Bunny in Disney World (a trademark impossibility). The key to implanting all these false memories was to repeat the questions over relatively long periods of time (by which point the volunteers would vaguely remember the information, but no longer remember where they'd heard it), and to invite the brain to imagine sensory detail ("Do you remember stroking Bugs Bunny's velvety ears?").

 Pioneering child psychologist Jean Piaget held a false memory about an attempted kidnapping when he was 2 years old. (Yes, wiseacre, he was the victim.) His memory included vivid visual details, such as the nearby subway station, the scratches his nurse received on her face while fending off the attacker, and the white baton of the policeman who interceded. Years later, the nurse confessed she had fabricated the whole story, and Piaget realized his memory had been created by hearing the story told and retold as a child. In a similar way, people may "remember" events that they see in childhood photographs.

Memories are influenced by mood

When you're depressed, you're more likely to remember the most miserable highlights of your existence on Earth. Furthermore, if you recall a memory while you're feeling utterly wretched, you'll describe it in far bleaker terms than if you recall it while you're in a decent mood.

People distort memories to fit the concepts they know

People rationalize memories to make more sense by dropping details that seem out of place, inserting new ones to fit the overall narrative, and reshaping everything else to fit. In a famous study, researchers tested volunteers on their ability to remember the "War of the Ghosts," a short Native American story containing supernatural elements that are particularly foreign to modern big-city dwellers. When urbanites recalled the story, they often dropped Native American details (for example, few remembered the protagonists were hunting for seals), while supernatural details were explained with elaborations (for example, "Something black came out of his mouth" was transformed to recollections of a man foaming at the mouth, or his soul coming out of his mouth). This transformational effect is similar to the one shown in the culturally ambiguous picture of the interacting family on page 88.

 The story-skewing effect shown in the "War of the Ghosts" study is thought to be the result of mapping new information into the existing framework of ideas in our brain. Ideas we aren't familiar with (for example, seal hunting) are more difficult to encode, while we rely on basic assumptions about narrative and causality to automatically fill in large amounts of most stories. You can find the full text for the "War of the Ghosts" at *http://penta.ufrgs.br/edu/telelab/2/war-of-t.htm*.

Easier to understand means easier to remember

The memory distortion effect hints at another, complementary principle. Quite simply, it's easiest to remember details that draw on the concepts you already know. When you encode this sort of memory, it will be more thoroughly connected to the rest of your brain.

For example, I have a good friend who's able to watch the same melodramatic TV movie a half dozen times without recalling the plot. However, she has an incredible ability to recall the complete lyrics of kitschy 80s songs and obscure commercial jingles from decades past. In part, her greater memory of this music is probably due to her paying better attention and repeating it to herself more often (two memory tricks you'll consider a bit later in this chapter). However, it's also likely that the extensive amount of musical knowledge in her brain (she's a practiced musician) has set up a framework of concepts that makes it easier to analyze a song and break it down into more easily remembered musical ingredients of key, harmony, and rhythm. A similar effect is found with chess grand masters, who are skilled at remembering arrangements of chess pieces on a board, presumably because they can chunk these arrangements into known positions and strategies.

Why We Forget

If pressed to remember the glories and tragedies of your own life, certain specific details will spring to your mind, vivid and alive. But countless more will remain just out of reach, slowly slipping into the gloom as the years advance, until the best you can do is attempt to reimagine the approximate outline of the moment you once inhabited.

Now is probably not a good time to tell you that your brain does you an invaluable service by forgetting things. But that seems to be exactly the case. Here are some reasons why a little bit of memory fog is often a good thing:

- **Avoiding information overload.** The brain is an amazing machine, but its capabilities aren't infinite. In order to make conclusions, create summaries, and see patterns, you need to be able to look past the details and focus on key themes. This task is more difficult if your mind is clouded with trivia.

- **Quick thinking.** In our evolutionary past, humans lived in a much riskier world. They relied on the ability to survey a scene and make quick decisions, preferably before being eaten by a giant bear. In this context, detailed recollections are more of a distraction than a help.

- **Assimilating new information.** The misinformation effect you learned about in the previous section, where new information is integrated into old memories, is a great strategy if you consider the brain's role as an all-purpose problem solver. However, it's not a good foundation for legal certainty, scientific investigation, and other areas where you need absolute certainty.

- **Avoiding emotional hangover.** The night after a drunken escapade involving a wedding toast, a cake topper, and the mother of the bride, you'll probably slink around in embarrassment. Fortunately, memories age like wines, and today's most cringe-inducing recollection will meld into a much more tolerable joke a few months later. One of the reasons we warp and reshape memories might be to deal with emotional disappointments, conflicts, and embarrassments in the healthiest way possible.

Although you'll never know what would happen if you had perfect recall, you can weigh these points by considering someone who *did*, as described in the next section.

The Man Who Couldn't Forget

Solomon Shereshevsky was a Russian journalist who had incredible powers of recollection. He could remember complex mathematical formulae he didn't understand, poems in foreign languages, and huge grids of numbers. Even more amazing was the fact that these memories became fused into his long-term memory. When tested years later, he could still remember the sequences of numbers he had learned, complete with unrelated details about what the questioner had been wearing, where they had been, and so on.

The basis of Solomon's incredible ability was *synesthesia*, a phenomenon where the experience of one sense (for example, vision) stimulates another (say, hearing). For example, someone with synesthesia might describe the number five as appearing purple. If this sounds a bit fishy, consider the synesthesia test shown on the next page (which is based on a test created by neuroscientists Vilayanur Ramachandran and Edward Hubbard).

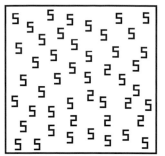 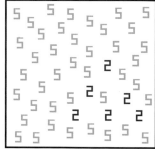

The triangle of twos in the right-hand picture is obvious for everyone (well except for red-green colorblind folks), but it doesn't jump out for most people in the left-hand picture. However, synesthetes can quickly pick out the triangle of twos there as well. For them, it stands out from the background of fives just as clearly as if it had been shown in a different color.

Solomon's extreme synesthesia led him to taste, smell, and see vivid images in conjunction with numbers and sounds. Carrying on a conversation while eating an apple would be unthinkable, because both activities would generate waves of conflicting impressions. Often, the synesthesia was unpleasant—for example, Solomon describes how after hearing the voice of the ice-cream seller (which gave him the vivid impression of black cinders bursting out of her mouth), he could no longer eat the ice cream.

 Note Although it's impossible to force yourself to experience synesthesia, you can use some of the same techniques that happen automatically in a synesthete's brain to boost your own memory power. For example, you'll learn in the next section how to tie numbers, words, and ideas to images and places, which makes them easier to recall.

Solomon's ability wasn't honed through practice—in fact, evidence suggests that he fought to forget. Because a single word could trigger a flood of memories and associations, he struggled to read a book or carry on a meaningful conversation. He also had trouble recognizing people if they wore new clothing or had unusual expressions, because the avalanche of unignorable details would overwhelm him. To learn more about his life, check out *The Mind of a Mnemonist* (Harvard University Press, 2006), a book by the psychologist who studied him.

Techniques for Better Remembering

If you're like most people, you've had no shortage of opportunities to insult friends and embarrass yourself with unexpected memory slips. Fortunately, there's a way to help. Although studies consistently show that you can't hone your memory simply by using it, you can improve your ability to store information by using a few odd tricks. Ordinary people who learn these techniques can quickly boost their otherwise miserable scores at simple memory tests, like remembering lists of numbers, names, and faces.

The art of improving memory is called *mnemonics* (pronounced NUH-moniks), and it's a time-honored practice that dates back to classical antiquity. In fact, it just might have started when an ancient Greek philosopher realized he couldn't find his horse and carriage in the Coliseum parking lot. All mnemonic tricks have to be applied at the moment you're actually storing your memories. They help you encode information in such a way that, later on, makes it easier to retrieve from the caverns of your brain.

Mnemonic tricks require a conscious effort, which means they're no help if you don't recognize important information when you see it. Similarly, mnemonic tricks can't help you remember something you've already forgotten.

 Note The question of whether memory decreases in old age (and if so, by how much) is surprisingly controversial. However, some studies, which found small but significant age-related recall deterioration, have also found that better learning strategies, like mnemonics, can more than compensate for the difference. To learn more about age-related brain decline and Alzheimer's disease, see page 237.

Paying Attention

The next time you're searching for your keys, grasping for a name, or lost in the mall, here's something to think about. The odds are that you haven't forgotten the information you need. Rather, it probably never entered your memory in the first place.

Studies consistently show that people don't bother to remember anything that doesn't scream out its importance. Consider some of the objects that decorate your daily life. Could you draw the pattern on your favorite coffee mug? Can you describe the clerk where you bought your last chocolate bar? Do you remember what your wife was wearing when you last saw her?

One crafty study showed that most of us can't identify an object we handle nearly every day—the lowly penny. Match it up with a few impostors (as in the lineup shown here) and most people are baffled.

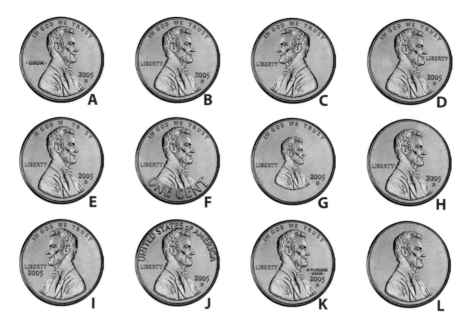

The human brain is a model of efficiency: Left to its own devices, it doesn't store details unless they're pumped up with emotion. This is a relatively good strategy. It ensures that you won't become drowned in a flood of meaningless trivia like Solomon Shereshevsky. After all, if you really want to know what a penny looks like, you can pull it out of your pocket and take a look. (It's coin B.)

But your brain isn't perfect, and left on its own it will happily ignore many details that are far more significant. Consider what happens to the average brain at a social gathering. After being introduced to a dozen or so new faces, the brain is all too happy to toss each name out of short-term memory only minutes later. At a friend's baby shower, it may be a welcome way to block out the endless parade of overbearing relatives. But in a business meeting, a forgotten name could send you sliding straight down the corporate ladder.

 Tip To pay better attention, stop multitasking. Your brain handles multiple tasks (for example, watching TV while studying organic chemistry) in much the same way as an average single-processor computer—it switches its attention back and forth incessantly. To stand a better chance of remembering, remove the distractions you don't want to retain so you can focus on what's most important.

The Practical Side of Brain Science
Remembering People at a Party

To forge a solid memory, you need to catch your brain before it starts to slack off. Here's how you can keep it on track in one of the most memory-challenging situations you're likely to face—meeting a large number of new people in a social situation:

- **Prepare.** If possible, get a list of guests who will be attending the party and familiarize yourself with it. When you actually meet the guests and put names to faces, they'll be easier to remember.

- **Identify the distractions.** And then ignore them. Top distractions include the setting, other conversations, and—most of all—yourself. If you catch yourself starting to wonder if your socks match or your breath is minty enough, put your mind back on track.

- **Force a look.** If you look away too quickly, you'll lose the chance to form a lasting mental image. Without that anchor, a person's name is meaningless. Instead, look at each new person you meet squarely in the eyes, and follow the next guideline.

- **Repeat new names.** A quick "Pleased to meet you, Madison" can help fuse a new name into your brain. It forces you to pay at least minimal attention, and gives you something else to try to remember—the sound of your own voice repeating the name. And if you don't catch a name, don't smile in embarrassment. Instead, follow up with a quick "I'm sorry, I missed your name." You may never get a second chance.

- **Convert the name to an image.** Pick a distinguishing feature, exaggerate it, and link it to the person's name or occupation. Silliness helps and painful puns are golden. For example, if you meet Rex, try to imagine him looking like a dinosaur. If you meet Mary, visualize a lamb at her side. If you see Miranda, conjure up a scene where she reads you your rights.

- **Be nice.** If someone seems to have forgotten your name, master the art of a knowing smile and a quick reintroduction.

Repetition

Paying attention in the first place is essential. To increase your odds of remembering the information you need later on, keep lavishing attention on the facts you want to remember using repetition.

This technique is simple, albeit a bit tedious. Simply force yourself to mentally review the details you're trying to remember. Along the way, ask leading questions about the significance of each item, how it relates to the whole, and what comes next. (For example, if you're trying to remember the identity of a person from your last company picnic, you might ask yourself a series of questions like this: "What's his name? What's his position in the company? Who else was he talking to? And how on Earth did he eat so many hotdogs?") The trick is to force yourself to spend more time mentally manipulating each image or concept, so it will be embedded more firmly into your brain.

Incidentally, repetition is also the basis for a definite non-memory-trick: *false memories*. False memories are memories that the brain embellishes or creates through a process of repetition and rehearsal. They're more likely in very young children (who often conflate fantasy and reality), people under the effects of drugs or hypnosis (which itself encourages wild flights of free association), and people influenced by leading questions from a therapist (which mimics the way a brain reconstructs a foggy memory by filling in uncertain details with conjecture). The end result is that instead of digging up a deeply repressed memory, the rememberer creates a fictional one.

Once created, a false memory is reinforced by constant repetition. By the end of the process, the false memory seems as real as any other distant recollection. In famous false memory cases, the memories often contain wild flights of fancy that become more elaborate the more the memory is rehearsed, such as sex with circus animals in physically impossible ways, Satan lunching on baby's brains, and so on. Unfortunately for the rememberer, dealing with a false memory is just as traumatic as dealing with the recovery of a genuine repressed memory.

The Method of Loci

The method of *loci* (places) is an ancient oratorical method that was all the rage among Greek orators in classical times. The idea is to mentally store memories in a well known location, like the rooms of a building. To understand the technique, it's best to try it yourself.

Whole Wheat Flour	Shampoo	Olive Oil	Battery
Cinnamon Sticks	Chocolate Milk	Rye Bread	Banana
Cauliflower	Key Lime Cheesecake		

The following grocery list is perfect for this test. Spend about a minute familiarizing yourself with these 10 items:

Then, put this book down for a moment, dig up a piece of paper, and see how many you can remember. (Really—it will help you compare the performance of the different memory tricks you'll learn in this chapter.)

Odds are you won't remember everything. The items in the grocery list aren't closely related to one another, and none of them are interesting enough to stick in your brain without help. Even if you do remember all the items, you'll need to use continuous concentration and repetition. The grocery list won't percolate down to your long-term memory storage, and when something more interesting captures your attention the entire list will be tossed out like last week's leftovers.

This is where the method of loci comes in quite handy. It's one of many mnemonic techniques you can use to help your brain file memories so they're easier to retrieve later on. To use the method of loci, begin by picking a location. (Your own home is a great choice.) Next, take a moment to imagine yourself walking through this location, traveling from room to room. Make note of *loci*—handy places where you could put an object (for example, in the stove, under the bed, on top of your mounted moose head, and so on).

Now, repeat the same journey, but take a moment to pause in each room and stash away an item from the grocery list. For example, the first item in the list is whole wheat flour. If you've chosen your house for your location, and you begin outside on the porch, you'd look around for a place to put your bag of flour—say, in the mailbox next to your door. Then, step inside and into the next room, and look for somewhere to put the cinnamon sticks. If you can think of combinations that produce striking imagery—for example, pouring the olive oil in your washing machine—you stand an even better chance of remembering it.

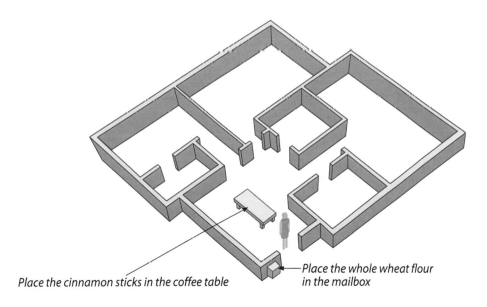

Place the cinnamon sticks in the coffee table

Place the whole wheat flour in the mailbox

This process of placing items in an imaginary location takes more time than simply reviewing the list. However, it has a big payoff—you're much more likely to remember the entire grocery list. Try it out with the same example, and see how much better you fare. The method of loci can work with lists of different lengths, but it lends itself particularly well to relatively long sequences (say, about 20 items).

 Note Legend has it that the method of loci was invented by a poet named Simonides in ancient Greece, after he stepped out during dinner just before the roof collapsed and killed the dignitaries inside. Simonides was able to list the dead by mentally walking through the seating arrangement.

Although you used a grocery list in this example, the technique works equally well with other types of lists. For example, if you have a list of people to phone you might imagine them lounging about the different rooms of your house. You can use variations on the theme to remember items to pack on a trip, ingredients in a recipe, your favorite books, your weekly goals, the reasons you deserve an immediate raise, and so on.

The method of loci has three winning features:

- **It invites you to visualize what you're trying to remember.** It's easier to remember the bag of whole wheat flour once you have a mental picture of it sitting in your mailbox.

- **It prompts you to remember each item.** This works because the method of loci binds something you know very well (your location) with something you're trying to learn (in this case, your grocery list). You're unlikely to forget the rooms in your house, or the path you took. But each room will give you additional cues that can help you remember the appropriate item. For example, if you know that you start outside on your porch, you'll quickly remember the mailbox, and then you need the slightest of mental jumps to remember what you put inside. This remembering is aided by the fact that you're the one doing both the storing and the remembering. In other words, if you're likely to think of putting something in a mailbox when you're reviewing the list, you're also likely to think of looking in the mailbox when you're trying to remember it.

- **It orders your items.** In the grocery list example, it doesn't matter whether the whole wheat flour is first or last in your list, as long as you get all the ingredients you want. But if you're using the method of loci to remember the logical arguments of a rambling speech (as were the ancient Greek orators), order is essential. This usage survives in a few common idioms of the English language, such as "in the first place" and "in the second place."

You'll find these three principles at work in the other memory techniques you'll learn in this chapter.

 The method of Loci is closely related to another memory technique called the *journey method*. The journey method asks that you visualize a journey (such as your daily commute to work, a favorite hike, and so on). You then place the items you need to remember at different landmarks along the way.

Stories

Another way to harness mental imagery to help you remember is by weaving the items in your list into a story. The trick is to use vivid imagery and create a narrative that's bizarre enough to force you to create some new neural connections.

Take the grocery list, for example. Right now it's just a series of random items, but with a feat of imagination you can transform it into the following tall tale:

> "A great snowstorm was falling. It looked as though the ground was covered in *whole wheat flour*. I walked in the snow, chewing on a *cinnamon stick*, until I saw a giant mound of snow that looked like an enormous head of *cauliflower*. I tried to scramble to the top, but I slipped on a puddle of *shampoo* and fell into a cold river of *chocolate milk*. Fortunately, a giant piece of *key lime cheesecake* floated past, and I climbed aboard. As I floated along, I snacked on a piece of *rye bread* slathered in *olive oil*. When I reached the end of the river, I saw the most amazing thing: a *battery*-powered *banana*."

Finding the images that mean something to you takes a bit of practice, but you'll get the fringe benefit of developing your inner Kafka.

 If you're serious-minded, you might not enjoy memorizing an admittedly nonsensical story. But studies conclusively agree that memory techniques like these will allow you to remember far more than you otherwise could.

If you're trying to remember a list of people, places, or concepts, you'll find that they're difficult to visualize with the story method. One technique that you can use to overcome this challenge is to rely on migraine-inducing puns. For example, if you're memorizing a list of Renaissance composers, you might visualize William Byrd flapping away, Orlando Lassus using a lasso to rope a wild horse, Claudio Monteverdi climbing a green mountain, Josquin Desprez talking to a dead president, and so on. The key is to choose images that make sense in the warped world of *your* brain.

Besides making sure the imagery is vivid enough to live on in your brain, you also need to make sure each item is *linked* to the previous one. You'll have just as much trouble remembering a sequence of unconnected images as the original grocery list. You can also chunk some items together into single images, as with the battery-powered banana in the previous example. If you get it right, you'll be amazed at how long your story can live on in your memory.

 Tip To create the most resilient memories, layer up with more than one memory-boosting technique. For example, the method of loci will work even better if you add a dash of narrative to explain why you're walking from room to room.

Word Games

Word games work because they ask you to remember something simple (a combination of cues, encoded as a word, sentence, or rhyme) rather than the raw information you really need. Here are three time-honored techniques that use word tricks:

- **Rhymes.** There are countless rhymes that encode simple but easily confused information. You may know rhymes for remembering the number of days in each month ("Thirty days has September..."), remembering a crude approximation of wonky English spelling rules ("*i* before *e*, except after *c*"), or even historical details (the fates of the six wives of Henry VIII are "divorced, beheaded, died / divorced, beheaded, survived").

- **Acronyms.** The frazzled students in technical fields love acronyms. One simple example from high-school math is PEMDAS, an acronym that indicates the order of operation for an equation (*P*arenthesis, *E*xponents, *M*ultiplication, *D*ivision, *A*ddition, *S*ubtraction). It works so well because the acronym is a single, speakable word. Even though it's meaningless, it can be stored in the brain as a single chunk of information. Bonus points for combining acronyms and stories: A Rat In Tom's House May Eat The Ice Cream (used to help at least one second grade class remember how to spell *arithmetic*).

- **Acrostics.** An acrostic is an acronym in reverse. For example, a common acrostic is the easily remembered phrase "*E*very *G*ood *B*oy *D*eserves *F*udge." If you take the first letter of each word, you have the musical notes that fall on the lines of the treble clef (E, G, B, D, F). Often, acrostics give letters that are linked to other words, which is particularly helpful if you need to sort out the order of known items. For example, the acrostic "My Very Eccentric Mother Just Served Us Nude" provides the order of our solar system's planets, without the recently demoted Pluto. For those in need of a refresher: Mercury, Venus, Earth, Mars, Jupiter, Saturn, Uranus, and Neptune.

Most people don't create word games; they just use a small set of time-tested ones that are passed down from generation to generation. However, there's no reason why you can't craft your own. In fact, if you expend the extra bit of effort, you'll find that home-cooked word games do just as well as memory aids. Use acronyms for short lists that can form speakable words (insert vowels if necessary). If you can't form a reasonable acronym, try a more complex acrostic. Rhymes, which don't always stick in the head as easily and can't capture information as concisely, should be your last resort.

 Note Wikipedia provides an eye-popping list of acronyms and acrostics, spanning fields from psychology to urology, at: *http://en.wikiquote.org/wiki/List_of_first-letter_mnemonics*.

Codes

When all else fails, if you can't remember the information you need then convert it to something else. This is particularly handy with a long series of apparently meaningless data, such as random numbers. It's also the technique that converts unremarkable phone numbers like 1-900-588-7867 into the much more memorable 1-900-LUV-PUMP through the encoding power of the telephone keypad.

In order for a coding system to work, you need to have the rules of the encoding system already stored in your brain (or readily available in a memory aid). Otherwise, the code isn't much help.

The most impressive mnemonic code system is the *major system*, which makes it possible to learn very long numbers by first converting the digits into sounds, grouping the sounds into words, and then weaving the words into the sentences of a story. Professional memory performers are very fond of the major system. They practice it until it's almost automatic, and then use it to remember seemingly endless lists of numbers, the order of an entire deck of cards, and so on.

Learning the major system takes some time, and it's really more suited to dazzling memory feats than everyday note taking. But if you want to give it a try, surf to *http://en.wikipedia.org/wiki/Mnemonic_major_system*.

Indexing Memories with Pegs

The way you store memories doesn't just determine whether they can be retrieved—it also determines *how* they're retrieved.

To see the difference, compare your performance at two similar puzzles. First, try to come up with a few animals whose names start with the letter T. Next, try to create a similar list of animals whose names end with T. Both lists include similarly common animals (compare *t*urtle, *t*iger, *t*urkey, and *t*oad versus rabbi*t*, ra*t*, elephan*t*, and ferre*t*). However, creating the first list is a relatively easy task while writing the second is a brain-stumping ordeal.

The difference is the way you navigate to the animal name information in your brain. When you learn to read, you spend a great deal of time sounding out new words by following their letters in a logical progression from left to right. When you learn new words, you incorporate them into this framework, indexing them based on the sounds they start with. From a neural point of view, you can follow the brain's connections from letters to sounds to animal names (and vice versa), but you won't find the same strong connections based on last letters.

 Note The brain's indexing system can easily adapt to different uses. For example, if you spend several hours a day trying to think of animal names that end in certain letters, you'll gradually get a bit better at it. Presumably, rappers and poets have a stronger set of connections that bring together rhyming words; similarly, Scrabble players can pick out the words they want to play by length and letter value.

With the memory techniques you've learned so far, you've indexed your memory by connecting it to a place, a journey, or a story. To remember a full list of items, you need to travel through the list in order. However, there's another memory trick that gives you more thorough indexing for a smaller set of items. It's called *pegging*, and it lets you jump to an item in any position without traveling through the whole list.

The most common peg system uses the numbers 1 through 10, and pairs each one with a vivid image based on a rhyme. For example, the number 9 is paired with *wine*. Here's the full list:

Peg Number	Peg Word	How to Remember an Item in this Position
1	Gun	Visualize the first item being fired from a gun.
2	Zoo	Visualize an association between the second item and a zoo (or your favorite zoo animal).
3	Tree	Visualize the third item growing from a tree.
4	Door	Visualize the fourth item behind a door.
5	Hive	Visualize the fifth item associated with a hive or with bees.
6	Bricks	Visualize the sixth item associated with bricks or a building.
7	Heaven	Visualize the seventh item associated with heaven or an angel.
8	Plate	Visualize the eighth item on a plate as if it is food.
9	Wine	Visualize a glass containing the ninth item.
10	Hen	Visualize the tenth item associated with a chicken.

Alternate options are possible. For example, 1 could be a bun, 10 could be a pen, and so on.

In order for the peg system to work, you have to become very familiar with these 10 images. Once you've committed them to memory, you can reuse them with any list by creating a vivid image. For example, with the grocery list you might visualize *whole wheat flour* being used as gunpowder (1), *cinnamon sticks* being eaten by the monkeys at the zoo (2), *cauliflower* growing from a tree (3), and so on. Now, you can remember individual items from the list by their numbered position, you can travel up and down the list, and you can quickly spot any missing items.

The obvious shortcoming of the peg system shown here is that it only accommodates 10 items. If you need to remember more, you'll be forced to double up, which increases the chances you'll forget something, or use a system with more pegs. Accomplished practitioners of this technique often peg items to the 26 letters of the alphabet or all the numbers from 1 to 100 (using various techniques to come up with the right images). You can find some of these peg systems on the Web.

Journals, Pictures, and Other Memory Aids

Sometimes the easiest way to remember something is to stop relying on the stunningly effective fact-dropping, key-losing, all-purpose forgetting machine in your head and just *write it down*. That way, you can get to work on the real job ahead of you (misplacing the piece of paper) and save a lot of trouble with imaginary places, Dadaist stories, and horrible puns.

This approach works well for a great many things, assuming paper is available, you have a logical place to put it, and you can capture everything you need to remember in a few words. The trick here is using a *memory aid*. In many cases, memory aids don't hold everything you need to remember; they're just a list of memory cues. One obvious example is written notes for a course. Although the best notes extract the key themes and facts, you still need to prompt yourself with leading questions to fill in all the details as you study. Other memory aids include appointment books, electronic organizers, and computer programs (like Microsoft Outlook).

Memory aids are particularly well suited to helping you recall the personal narrative of your life. Here, the challenge is to remember a collection of details over large distances of time. A carefully chosen memory aid—such as the entries in a journal, photographs, or video recordings—can become an anchoring point that helps you recall a wealth of related information. (Kitschy souvenirs aren't recommended.)

 Tip Many people find that full-fledged journals are just too much of a chore to keep up. However, you can create similar anchoring points without recording the full detail of a journal entry. One good example is a *themed list*, such as funny things said by your preschooler, tasty meals at restaurants, recently read books, and so on. Maintaining these lists is easy, and they often provide just enough of a jumping-off point for your memory to fill in the complete picture.

One side-effect of keeping these sorts of records (particularly photographs and video footage), is that over time the information that's linked to your memory aid is strengthened, while the information that's not becomes forgotten.

For example, imagine you take pictures of a three-week odyssey through Siberia, but your camera froze in Yekaterinburg. Over the years, as you review the pictures you took, describe them to family, arrange them in an album, and so on, you'll practice recollecting your happy Siberian moments. However, your days in Yekaterinburg won't have the benefit of this repetition process. Over time, you'll probably remember much less of that portion of your trip. The same distinction occurs between sights and activities you photograph and those you omit.

Gimmicks

Sometimes it's the simplest tricks that make the difference between timely remembering and a forehead slap. Here are a few ideas that might increase your remembering odds:

- **Speak aloud.** If you know you're about to do something you might forget later (for example, putting your keys down in a fishbowl), say "I'm putting my keys in a fishbowl." This technique has the side-effect of amusing anyone who's around you and it might make you think twice about a half-witted action.

- **Put reminders in unmissable places.** Need to deliver a few letters on the way to work? Don't pop them in your bag, where they're sure to be over-looked. Instead, slip one into your shoe so you can't possibly leave your house without remembering what to do. Afraid you'll forget your carefully prepared lunch once again? Put a post-it-note at the very last place you'll look before you exit your home—say, the door on the way out.

- **Cultivate odd tricks.** For example, did you know you can figure out whether a month is short or long using the knuckles of one hand? Start at your index finger (the one next to your thumb) and head to your pinky, placing a month on each knuckle and valley in between. When you reach the pinky (July), start over again at the index finger. Every month on a knuckle is long (31 days), while every month in a valley is short (30 days, except for February, which is 28 or 29).

- **Put it to music.** This trick has more mileage for some people than others. However, happy hummers have been known to set entire mathematical proofs to the tune of hit pop songs. It may be the closest Britney will ever get to calculus.

- **When all else fails, carry paper (or a digital organizer).** After all, do you really want to go around all day with a bizarre image of a lion giving birth to an elk because you parked your car in section LBE?

Better Learning

Remembering is just one part of learning. And while remembering is a clearly defined challenge, learning is a lot more subtle. It depends on intangible things like personal experience, sadomasochistic instructors, field trips, and knowing when to listen and when to ask annoying questions.

That said, the information you've gleaned in this chapter and the previous ones provides a few useful insights into what learning strategies might work (and what ones are obviously doomed to fail). Here's a brief list that consolidates a few decades of educational research:

- **Use multiple modalities.** Most people have a preferred way to learn. Some master new facts by hearing them and copying them down, others rely on visual aids and imagery, while still others need to put the information into practice. If you're a student, determine how you learn best and try to devise practice sessions accordingly (rather than emulating other people's study habits). If you're a teacher, try to include a rich range of materials and exercises that invite students to engage with the materials in a variety of ways.

- **Attention needs engagement.** Studies show that so-called sleep learning—a dubious practice whereby you listen to recorded lectures while snoozing—is a wash out. The only way to really assimilate information is to pay attention. One of the best ways of encouraging attention is through questions. If you're an instructor, use the infamous double-questioning tactic. Ask students open-ended questions (those without *yes* or *no* answers) to keep them on their toes, and then force them to come up with their own follow-up questions later on.

- **Make it yours.** The point where learning occurs is often the point when you connect rote memorization to your own framework of ideas and experience. Once you've got a solid handle on the key facts you need to learn, it's time to manipulate it in your brain. The best approach depends on the subject matter and the learner, but you can try grappling with your newfound knowledge in conversation, by writing out new summaries, practicing with sample problems, case studies, and role playing, reading secondary sources that force you to reexamine what you've learned, and so on. The more you turn concepts over in your mind's eye, the more connections you'll be able to forge to the facts you already know, and the better you'll be able to put your knowledge into practice.

- **Teach it to someone else.** This technique works on multiple levels. It forces you to reorganize the material in your head, repeat it aloud, elaborate it, and answer questions, all of which help cement the knowledge even more firmly into your own brain.

- **Take frequent breaks.** A simple walk around the block can give your brain some much needed time to incorporate new concepts. Also, think twice about how much learning you stuff into a single day. As you learned in Chapter 4, the brain adjusts to constant unchanging stimulus by tuning it out. A classic experiment in learning compared students who tackled a new subject in two 2-hour intervals a day against those who had just a single 1-hour interval. The heavy-learning group did master the subject faster (over fewer days), but altogether they invested twice as many learning hours. So mad cramming will get you there, but it's not the most efficient use of your time.

- **Follow up.** The brain is a relentless garbage collector. Even after you've mastered a new subject, your knowledge will grow soft and flabby if you don't draw on it. If you aren't currently using some knowledge that you want to retain, give yourself a periodic refresher.

- **Have the right attitude.** If you're a teacher, have high expectations— it's more likely to convince fence-sitters that it's worth putting in the effort. If you're a student, cultivate an attitude of life-long learning. Realize that the most successful people in life openly admit their islands of ignorance, and are always venturing out into new fields of learning.

6 Emotions

In the early days of human civilization, the brain was (somewhat humiliatingly) overlooked. Despite a few physicians and philosophers who were on the right track, most people thought the *heart* was the seat of thought, morality, and intelligence. Aristotle suggested that the brain was nothing more than a portable radiator designed to cool blood. The Bible failed to mention the brain at all, instead stressing the three organs that Hebrew thinkers thought were most important to the human soul—namely, the heart, kidneys, and bowels (leading to charming turns of phrase such as "My kidneys shall rejoice" [Proverbs 23:16] and "My bowels are troubled for him" [Jeremiah 31:20]). To this day, the English language still bears the marks of this age-old heart obsession. After all, when was the last time you had a *brain-to-brain* with your significant other, described baby kittens as *brain-warming*, or implored an unfeeling cynic to *have a brain*?

Although our language is rooted in the past, today's science recognizes that the brain is the center stage for emotion. If there is a competition between an intellectual calculating machine and an emotional core, it all goes down in the billions of neurons in the brain.

In this chapter, you'll learn why you have emotions, how they work, and why the third piece of chocolate cake rarely tastes as good as the first. You'll tiptoe through the minefield of chronic stress and hunt for the ever-elusive state of happiness. By chapter's end, you'll have the distinct impression that your brain is running yet another part of your life without letting you in on the deal.

Understanding Emotion

Scientists have had a surprisingly hard time agreeing on exactly what emotions are. In the early days of psychology, there was a tremendous debate between those who thought emotions were the brain's way of firing up the body ("You have offended me. Now I will become angry.") and those who had it the other way around, and believed emotions were the brain's interpretation of the body's changing states ("I feel strange. This must be anger."). Now, scientists have fought their way to a truce that lies somewhere in between. Here's what they think.

Your Built-in Emotional Programming

From birth, your brain comes pre-wired with a few key emotional responses, like pleasure and fear. These emotional responses are a fundamental part of the human condition—in other words, even someone from the most isolated tribe in the South Pacific has exactly the same emotional programming as you do. They might describe their emotions differently, and they might *apply* them differently (for example, they might not find much to fear in a family of cockroaches, or they might devoutly worship our discarded bottles of Coca-Cola) but they'll still experience the full gamut of human emotion in their quite different lives.

Some of the best evidence supporting this argument is found in cross-cultural studies comparing human facial expressions. When shown pictures of people from a remote culture, we have no trouble identifying surprise, anger, fear, disgust, grief, and the rest of the lexicon of human emotion. Similarly, although those South Pacificans can't make sense of our laptops, smart phones, and iPods, they have no trouble interpreting the look on our face when one of these devices conks out on a Micronesian island miles from civilization.

Note Facial expressions are an instinctive form of communication. Before humans developed language and sunglasses, they probably spent a lot of time staring into each other's faces to learn about nearby threats and gauge the sincerity of their companions. Although modern humans carefully control their expressions, this control isn't absolute—just consider the noticeably phony smile of the car salesman or the guilty look on the co-worker who snagged your last Oreo. So if you want some insight into what other people are really feeling, it's worth cultivating the art of face watching.

Facial expressions are only one line of evidence that shows we share a common library of feelings. Another clue is the fact that we rarely encounter emotions we can't understand. Although you might have serious trouble unraveling the Old English in the epic poem *Beowulf* (which is at least a thousand years old), you won't have any difficulty interpreting its themes of kinship, loss, envy, and revenge.

Our basic emotions are biological drives that have been shaped over millions of years of evolution. The brain uses them to keep us on the right path—in other words, out of the way of dangerous beasties (fear), away from decaying produce and fecal matter (disgust), in control of our precious resources (anger), and in hot pursuit of a good meal and a good mate (pleasure and lust). These emotional responses also activate our bodies in specific ways. If we see a potential predator, we freeze in place, our heart beats faster, our lungs breathe faster, and our blood is diverted from our skin to our major organs. If we see a potential mate… well, you get the idea.

What an Emotion Feels Like

As you learned in Chapter 1, there are higher roads and lower passages in our brain. Emotional drives start in the lower passages. They happen *to* us, and are automatic, involuntary, and almost always inconvenient.

However, the higher level areas of our brain—the deep thinking cerebral cortex—*perceives* these emotions even as we *experience* them. And there's good reason to suspect that this combined experience of having an emotion and simultaneously reflecting on it is a different cup of tea altogether. In fact, it probably makes the difference between the way a snake swallows mice and the way we eye a triple-chocolate brownie.

Note Some neuroscientists give other neuroscientists headaches by carefully distinguishing between the words emotion and feeling. They use *emotion* to describe the brain's auto-programmed response to certain stimuli, and *feeling* to describe our conscious impression of that response. For example, when confronted with a peckish polar bear our brain launches into a defense program that gets us ready to run. We interpret this brain state as a feeling of fear.

Many of the things that we consider emotions might not be simple emotional drives, but more complex mental states that are hard to pin down. For example, the affection you feel for a romantic partner is probably part of an emotional drive that's designed to keep you together in a partnership long enough to send your genes careening into the future. However, the fact that this hodgepodge of lust and bonding metamorphoses into the social force called *love* probably has more to do with the influence exerted by the brain's higher thinking centers. (Certainly we don't see baboons exchanging flower bouquets, coveting chocolate bunnies, and singing in Bee Gees falsetto.) A similar process of transformation works with other more subtle emotions—for example, converting ordinary fear into shyness or desire into jealousy.

How the Brain Assesses Emotions

When your brain is interpreting how an emotion *feels*, it takes physical cues into account. In one famous experiment, volunteers were given adrenaline (a hormone that stimulates the body in the same way as many emotions). If left alone, these volunteers reported that they didn't experience any particular emotion. But if left with a buffoon or questioned about a painful event, they described the experience as being a whole lot funnier or a whole lot more upsetting than people who didn't get the adrenaline injection. In other words, the adrenaline-dosed volunteers subconsciously noticed their faster heart rates and higher state of arousal, interpreting both as signs that they were experiencing a stronger emotion.

Other studies have suggested it's not the physical cues, but our *perception* of those cues. One famous experiment made people feel that they were experiencing stronger emotions by playing a tape of a speeding heart beat while showing them pictures of scandalously attractive men and women. The trick was that the people were told that the increasing heart rate they heard was their own. A complementary line of research tells us that people with spinal injuries rate their emotions as feeling weaker, depending upon how much sensation they've lost from their body.

Note Of course, body cues don't determine the emotion you feel. They simply influence its perceived intensity. After all, the same state of sweaty-palmed arousal can underlie a high-stakes sporting event or a narrowly missed car accident.

Are We Programmed for Aggression?

In any discussion about emotional drives, the same question eventually comes up— namely, how much control do we have over our deep, dark, instinctive animal natures? And can our animal ancestry explain stealing, cheating, brawling, and The Jerry Springer Show?

Aggression is a perfect example. It gave our ancestors an edge in survival as they competed for food, mates, and territory. However, in the modern world acts of violence are often senseless and counter-productive. Your boss isn't more likely to promote you if you brandish a baseball bat. Disabled vending machines won't spring to life when you kick them. The daft drivers in front of your car won't clear out of the way when you scream a few colorful expressions.

At first glance, this seems like another case of the old brain, new world problem. But the truth isn't that clear cut. Our huge cerebral cortex (page 10) makes us the world's most flexible species, with learning abilities that dwarf our inherited instincts. So while a cat just can't help but pounce on that temptingly helpless bird, we have plenty of chances to overrule our impulses.

And even if we sometimes do succumb to our baser instincts, no one can agree about exactly *what* those instincts are. For example, there's no doubt that aggression helped our ancestors fight and conquer. However, many anthropologists think affection and altruism played greater roles in keeping us alive, allowing us to band together to solve problems, face danger, and raise children. In other words, at the same time that evolution was training us to be aggressive, it was also shaping us into compassionate nurturers and good team players.

Even our animal relatives don't shed much light on the issue. For years, evolutionary psychologists were obsessed with our nearest neighbor, the common chimpanzee, who lives in an occasionally violent tribal society ruled by an alpha male. However, another species of closely related chimpanzee—the *bonobo*—offers a dramatically different example. Bonobo society is ruled by groups of females, and social conflicts are resolved with rampant sex instead of violence. In fact, bonobos use impromptu sex to greet each other, trade favors, diffuse tension, and cement the bonds between the ruling caste (yes, this means girl-on-girl action between the dominant females). So even if you look to the animal kingdom, it's up to you to decide whether we're genetically programmed for tribal dominance or sexual dalliance.

The brain doesn't just evaluate the strength of an emotion, it also makes guesses about its cause, and these assumptions are just as easily misled. For example, when experimenters stimulate a part of the brain that makes people laugh, the experimental subjects are quick to fill in the blanks about why they're laughing. One girl under surgery to cure her epileptic seizures described it this way: "You guys are just so funny...standing around." As you'll see in Chapter 7, the brain is one big machine for generating over-simplified conclusions. One of the things it loves to explain is emotion. Unfortunately, it's rarely correct.

So what does this tell us? Old standbys like "follow your heart" probably aren't good advice, because your brain's emotional centers are fickle, un-predictable, and thoroughly beyond your control. They're better at getting your immediate needs settled than helping you chart out life decisions. In short, you should heed your emotions the same way you listen to physiologi-cal signs like hunger and fatigue. After all, they tell you quite a bit about what's going on in your body. Just don't expect emotions to give you the final word on a challenging issue, and don't assume they'll send you the right way when you're deciding whether to change job, homes, or life partners.

Pleasure: The Reward System

Quick, what do sex, a job promotion, and the act of defecation all have in common? Shortly after the act is complete, your brain rewards you with a brief flash of pleasure to let you know your life is on track.

Pleasure is the brain's reward system. It encourages you to pursue the ac-tivities that are in your biological best interests—activities that keep you healthy, well fed, and in top procreating form. Pleasure also greases the wheels of social interaction, helping you form lasting alliances with your own kind.

The *nucleus accumbens* is the leading candidate for the brain's pleasure center. When rats were given the chance to electrically stimulate this area in their brains, they hit the lever thousands of times an hour, showed no interest in food or mates, and eventually died of exhaustion. In other words, you may think you love sex, money, and chocolate cake, but what you really want is something a whole lot better: the tiny current of electricity that your brain uses to reward you.

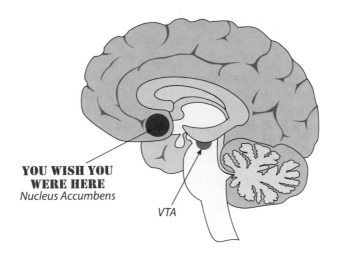

**YOU WISH YOU
WERE HERE**
Nucleus Accumbens

VTA

The nucleus accumbens may be the part of your brain that generates the feeling of pleasure, but it doesn't work on its own. A region called the *ventral tegmental area* (VTA) sits at the very core of your brain and receives all sorts of information that indicates how well you're doing in satisfying your biological needs. It then tells the nucleus accumbens to dispense a little pleasure for a job well done.

 Note The primary (but not the only) way that the nucleus accumbens communicates with the VTA is by releasing a neurotransmitter called *dopamine*.

Running out of Pleasure

Now for the bad news. Not only is the brain designed to give you pleasure, it's also designed to hold it back. Here's why:

- **Pleasure is only an effective motivator when it's in short supply.** If a simple piece of cheesecake gave you waves of pleasure that lasted for hours, you wouldn't need to eat for the rest of the day. Instead, the pleasure dies off quickly, sending many of us back to the fridge for another piece. This isn't all bad—after all, if your brain was more generous with pleasure, you'd have trouble getting motivated for tasks that take time, like learning to play an instrument, starting a new business, or writing a book.

- **There's a lot your brain wants you to do.** Eating and procreation are pleasurable, but you also need both to stay alive and spread your genes. To make sure you engage in all of life's important activities, your brain quickly grows accustomed to new sources of pleasure. That means you won't get the same effect from a second or third piece of cheesecake that you got from the first. (There is an exception to this rule. If you suffer from emotional eating—described on page 42—you may eat compulsively to relax, to distract yourself, or to deaden other emotions.) Similarly, after a sexual romp you'll be ready to engage in a different brain-pleasing behavior, like getting a snack.

The brain is continually calibrating itself to your current experiences. It dishes out just enough pleasure to get you to meet your body's requirements. In fact, it's quite possible that the pleasure you get from a fudge cookie is no more than the pleasure our prehistoric ancestors had when chowing down on a decidedly less appetizing piece of dried tree squirrel. Either way, the brain recognizes that you're satisfying your need for calories with the richest food available, and so it thanks you.

The Practical Side of Brain Science

Prolonging the Good Feelings

Your brain adjusts itself to prolonged pleasure in the same way that it adjusts itself to any repeated stimulus—it starts ignoring it. You can easily test this effect with a big box of your favorite Belgian chocolates. The first will be heavenly, the second will be pleasant, but the last one won't be much more satisfying than chewing a piece of plastic fruit. Why does your brain have to be so cruel?

Once you understand how the brain runs your life with its pleasure circuit, there are a few good tricks you can use:

- **Spread out the pleasure.** If you want every chocolate to be as mind-blowing as the first, don't eat them all at once. A short delay helps, but the best solution is to ration them over several days. It sounds a little controlling, but it delivers the maximum pleasure per cocoa-calorie. This sort of disciplined self-deprivation may be behind the legendary French attitude toward savoring food (and staying skinny).

- **Switch from one type of pleasure to another.** If you want to keep yourself in a state of ecstasy, switch from one neural pathway to another. Start with chocolates, then listen to your favorite music, get out and enjoy a sunset, hug your lover—you get the idea. Who knew that leading a life of sensual pleasure was so much work?

- **Don't expect too much.** Expect your pleasure to be short lived. That way, you won't be disappointed, and you might save a few dollars by avoiding new toys and four-star restaurants that won't keep you happy for long.

Motivation

In many respects, pleasure is the end of the story. After all, pleasure is a reward for completing a task. But what gives you the motivation to start out a task in the first place? Although it's a contentious issue, many neuroscientists have tracked the source of motivation to the *prefrontal cortex* (PFC), a critically important area that sits at the very front of your brain.

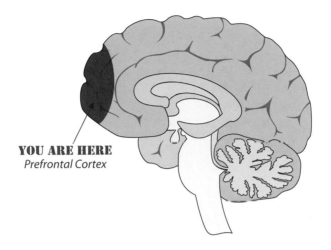

YOU ARE HERE
Prefrontal Cortex

As you'll see in the next chapter, the prefrontal cortex is linked to everything from personality to good planning. Among its many responsibilities is *motivation*—in other words, compelling you to pursue rewarding goals.

Researchers have discovered the link between motivation and the PFC by studying people with brain damage in this area. Although the effect of this damage depends on exactly what gets injured, many sufferers lose the ability to visualize the future. They can carry on intelligent conversations, enjoy a fine dinner, but are spectacularly unable to imagine what they'll be doing a few minutes from the present and completely unable to plan for it. Some are able to enjoy a reward but don't experience the desire to go and get it.

The Many Ways to Please You

You can obey your brain's reward system, you can fool it occasionally, but you can never escape the fact that your brain gets the final say about what's fun and what's not. However, before you conclude that you're doomed to become a mindless automaton in search of bodily pleasure, there's something you should know. You have the power to influence what activities your brain thinks are worthwhile.

Remarkably, your brain's pleasure center responds to less tangible rewards than food, sex, and comfort. For example, studies show that the nucleus accumbens lights up when you're expecting to win some money (and it's also implicated in addictive behaviors from gambling to video game playing). This shows that the brain doesn't just reward you when you satisfy obvious physical needs; it also rewards you for activities that your conscious brain *believes* are particularly beneficial. The brain needs this wiggle room to let you pursue important long-term goals, like finding a romantic partner, building a home, and raising a child. It also means you can derive some pleasure from activities that wouldn't have meant anything to our cave-dwelling ancestors, like solving a math problem or gaining social prestige with your killer dance moves.

It's likely that your prefrontal cortex plays a key role in allowing you to derive pleasure from secondary goals. Essentially, it draws on your memory of emotional experience to imagine the potential reward of different activities. (For example, money = new shoes = attention from opposite sex = opportunity for a reward.) For this reason, many people find that acquiring bits of unremarkable green paper can be highly pleasurable. But be careful—your prefrontal cortex is capable of reality-straining leaps and overstretched associations, which is why you find yourself plunking down twice as much for the brand name aftershave lotion with the sexy man on the package.

Cheating the Reward System

Humans have worked long and hard to circumvent the brain's stingy reward system. Some of the techniques we've used are clever, others are poor substitutes, and a few are downright dangerous. Here are some thought-provoking examples:

- **Birth control.** This successful dodge gives humans all of the pleasure with none of the babies. (And many people go one step further to get most of the pleasure with none of the company.) Like fundamentalist ministers, the brain can't distinguish between sex and procreation, so it dishes out its reward even when you aren't creating the next generation.

- **Adoption.** This noble enterprise gets parents all the same warm feelings, with none of the gene spreading. In our prehistoric past, adoption was probably an all-in-the-family affair, where childless adults would look after nieces, nephews, and cousins. But the brain has no kin identification device, so it's happy to bolster bonding with any cute bundle of joy.

- **Artificial sweeteners.** Modern chemistry has created a slew of look-alike compounds that seem like sugar and fat on our tongue, but can't be digested in the same way. A more dangerous example of fooling the food-reward system is found in sufferers of bulimia, who eat their food and them empty it out.

- **Alcohol and drugs.** One of the most dramatic ways that humans try to cheat our brain's stingy reward system is with drugs. Different drugs work in different ways, but most exert some sort of influence on the brain's pleasure circuit. (Often, they change the level of neurotransmitters like dopamine, which the VTA uses to tell the nucleus accumbens to give you some pleasure.) However, the brain uses its self-calibration system to adjust itself to these changes, even going as far as to kill off the very cells that create the intoxicating drug high. The result is well documented: repeat drug use becomes less and less pleasurable, life without drugs becomes more and more miserable, and drug addicts lose their ability to motivate themselves and enjoy previously rewarding things. In effect, their brains have become so insensitive to the neurotransmitters that signal pleasure that they hardly even notice them anymore.

Fear: Avoiding Death

Pleasure isn't the only tool at your brain's command. Its obvious complement is pain, which alerts your brain when you've crushed a toe or broken a tooth. However, most of us think of pain as part of perception rather than an emotion. (For the moment, we're ignoring psychological states that might be described as painful, such as sorrow, grief, and despair.) And though the line is a bit blurry, pain kicks in at a lower level. Your body has specialized neurons that perceive different types of discomfort and notify your brain about the problem. There isn't much room to wiggle out of it.

A more interesting comparison is between the motivation of desire, which pulls us toward certain things, and fear, which pushes us away. Much as the brain has a sophisticated pleasure circuit to reward good deeds, it also has an intricate fear circuit for reacting to potential dangers. It's almost like the brain is a very old school parent, bribing us into the right behaviors and smacking our mental bottoms to get us out of harm's way.

The fear circuit is rooted in two small almond-shaped brain regions called the *amygdala* (there's one in the left side of the brain, and one in the right side). The amygdala is buried deep in the brain, underneath the pleasure circuit.

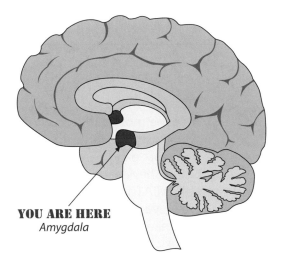

YOU ARE HERE
Amygdala

 Note Pleasure and fear are the two fundamental poles of the brain's emotional self-preservation system. Other emotions, like anger, affection, and disgust, involve the pleasure circuit and the amygdala, and they probably also tie in other brain structures that haven't been explored as thoroughly.

The Fight or Flight Response

In Chapter 4, you learned about the curious phenomenon of *blindsight* (page 73), where part of the brain is able to react to something even though it isn't consciously perceived. Blindsight shows that there's more than one pathway in your brain that responds to the things you see and hear.

Fun Facts

Rock Out Like a Neuroscientist

The amygdala is one of the few brain structures to inspire the name of a rock band. The *Amygdaloids* is a surprisingly tight outfit fronted by leading neuroscientist Joseph LeDoux. Their memorable music combines classic rock stylings (they call it heavy *mental*) and inventive lyrics that grapple with the realities of 21st-century brain research. Popular songs include "Mind Body Problem" ("My body wants you so, but my mind just says no") and "Memory Pill" (based on LeDoux's ground-breaking study in which he erased a single memory in the brain of a rat). To learn more or preview their songs, visit *www.cns.nyu.edu/ledoux/amygdaloids*.

For example, imagine that you're in a clearing in the jungle looking at a snake that has suddenly appeared in front of you. To consciously perceive the snake, the information that's registered by your eyes takes a relatively leisurely jog to the higher processing centers in your visual cortex. At the same time, more limited information is funneled down an older passageway to the amygdala.

If you didn't have the benefit of this pathway to the amygdala, the exchange might go something like this:

> Ah, what's that then? Gray, sort of stringy shape. Familiar that—I've got it! That's a snake. Is it coming this way? Oh, biting me now. Oh dear, sharp teeth. Feeling a bit queasy. Help!

With the benefit of your amygdala, you process the snake's appearance unconsciously and automatically. The exchange is more like this:

> Fast movement. Me no like. Back away!

In short, the amygdala gets the bare information it needs to trigger a life-saving response, before you have the chance to think about what you're looking at. This strategy will result in quite a few false alarms (for example, it might react if you see a snake-shaped garden hose ahead), but it also might save you from meeting your end in the jaws of a black mamba.

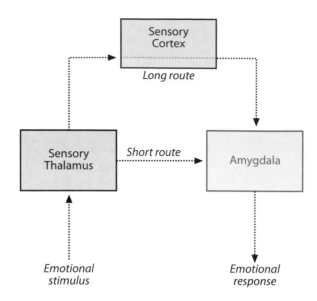

Long route

Sensory
Cortex

Sensory
Thalamus

Short route

Amygdala

Emotional
stimulus

Emotional
response

When the amygdala decides there's danger ahead, it communicates with the **hypothalamus**, the brain's built-in hormone-dispensing drugstore. With the help of the hypothalamus and other parts of your endocrine system (page 20), your brain floods your blood with hormones like adrenaline and cortisol to get your body pumped up and ready for emergency action. The amygdala also interacts with the low-level brain regions that control movement. Depending on the amygdala's snap judgment, you may run madly away, assume a defensive posture, or (most commonly) freeze in place.

Freezing is one of our most basic survival responses. It keeps us from moving toward the danger, gives us a moment to size up the situation, and possibly prevents a threatening creature from spotting us. In the meantime, the body is racing forward under the effects of fight-or-flight hormones so that when you *do* react, it will be with a desperate burst of energy.

 Often, the amygdala will sound a false alarm. It's then up to your higher-reasoning cortex to switch it off.

Emotional Memories

Some things are instinctively frightening, such as sudden movement, loud noises, and sudden changes in lighting. However, there are many more things we need to learn to fear. If your heart starts to pound when you receive your mortgage statement in the mail, you've noticed this phenomenon.

The anatomy of your brain explains why this happens. Your amygdala receives information from several other areas of your brain. It not only pays attention to what you see and hear, but it also retrieves vital information from your memory.

Unfortunately, when something traumatic happens the amygdala urges your brain to store a whole whack of details about the event, including some that obviously aren't related. For example, if you're mugged at night on Broadway by a Chinese woman, you'll have instantly acquired some sensible and some unrealistic associations. Not only will you become afraid the next time you're walking alone in the dark (a sensible reaction), you might also tremble when you set foot on Broadway with friends during the day, and you might break in a cold sweat when you run into a Chinese co-worker in the parking lot. In other words, the highly associative emotional reactions of the amygdala can underpin many a senseless prejudice. After all, the amygdala's job is to save your life—and in the evolutionary scheme of things, a few daft ideas are a small price to pay to avoid winding up becoming a carnivore's lunch.

The amygdala also influences memory consolidation—in other words, it helps determine whether you'll store a particular detail in long-term memory. An excited amygdala also helps you create *flashbulb memories*, vivid recollections that include the minute detail of an entire scene. It's thanks to flashbulb memories that people can answer questions like "Where were you when you heard about the terrorist attacks of September 11?"

The Practical Side of Brain Science

Controlling Your Emotions

The obvious difference between humans and many less impressive animals is that we don't need to spend our entire lives on emotional autopilot. We can, and often do, override our instinctive emotional responses to respond to changing situations.

However, there are a few caveats. The connections that funnel information from the amygdala to the cortex are stronger than those that funnel information from the cortex to the amygdala (which you use when you want to exert conscious control over your emotions). This means it's easier to *turn on* the fight-or-flight reaction than to *turn it off*, which is part of the reason we're so easily afflicted by chronic stress.

Children are in a particularly difficult situation. The amygdala is mature at birth, but the pathways that connect the cortex to the amygdala aren't as developed. This immature wiring just might be to blame for raging toddler tantrums.

Stress

In prehistoric times, the fight-or-flight response prepared people for the only actions they had at their disposal. But in the modern world, where we're more often engaged in mental challenges and it's considered bad form to stab someone who corrects your grammar, the fight-or-flight response isn't always appropriate.

Mild levels of arousal can improve attention and performance in school exams, sporting events, and heated debates—because all these actions are short and allow you to respond. Problems occur when you face a stressful situation that lingers on and doesn't provide an obvious avenue for you to *act*. For example, if you're trapped under the thumb of a sociopathic boss in a dead-end job, but you won't have money to pay next month's rent if you leave, you're in trouble. The constant stress of the situation will continually ramp up your body's fight-or-flight response, while you struggle to continually inhibit your body's natural instincts. After months and years of a situation like this, your body won't be the same.

The Effects of Stress

Constant stress is like having a car alarm going off in your body around the clock. Eventually, you'll learn to tune out the cacophony. However, you'll still end up with a wicked headache at the end of the day.

When your brain feels threatened for long periods of time, your body experiences the following changes:

- **High blood pressure.** The fight-or-flight hormones that get your body ready to act will also eventually wear it out. The list of potential complications from high blood pressure is long, including heart, eye, and kidney damage.

- **Faltering brain and memory.** The fight-or-flight hormones also weaken your ability to concentrate and form new memories. In fact, several studies suggest that the *hippocampus* (the brain structure that's responsible for long-term memory storage) begins to shrivel up under the influence of constant stress.

- **Disease.** As part of the flight-or-fight response, your body releases glucose into your blood to provide more energy for serious athletic feats (like sprinting away from a bear). But over the long term, high levels of glucose can damage cells throughout your body and aggravate diabetes.

- **Weakened immune system.** Cortisol and other stress hormones have a natural anti-inflammatory effect, which prepares you to deal with injuries. However, they also weaken the immune system, making stressed people more susceptible to infections.

- **Weight gain.** As you learned in Chapter 2, cortisol promotes weight gain. And because the fight-or-flight repsonse diverts blood from your intestinal organs to your muscles, you'll have a harder time digesting your food.

- **Dampened sex drive.** Long-term stress also decreases sex drive in men and women, possibly through a reduction in testosterone.

All of these problems have the same cause. The fight-or-flight response diverts energy from tasks like digesting your lunch, maintaining your body, and spreading your genes, which are luxuries when the brain believes you're in a life-threatening situation.

Conquering Stress

A short-term bout of stress is harmless. The real problem is *chronic* stress, which is sustained over months or years. If chronic stress is consuming your life, or if you just want to make sure it stays away, heed these stress-shrinking tips:

- **Relax!** It's not as difficult as it sounds. Although everyone has different ways to tune out and gear down, you probably already know yours. It doesn't matter whether you use music, meditation, or naked shamanic drum circles—make the time to practice your personal stress relievers.

- **Exercise.** Aerobic activity—exercise that forces your heart to beat faster—is a particularly good tool for battling stress. It gives your body the chance to respond in a physical way, without engendering any lawsuits

- **Improve your environment.** The modern clutter of sounds, sights, and smells can keep you on edge. Avoid living in places that are submerged in a din of traffic noise or punctuated by the late-night shout-outs of nearby neighbors. Aim not just to reduce sound, but to clear out *information*, which includes everything from busy posters to disordered desks. Television is a particularly insidious villain—it's perfected the art of continuous attention grabbing.

- **Mentally reframe the situation.** Stress is the result of what goes on in your brain, not what happens in your environment. If you picture an obstacle as a worthy challenge rather than an infuriating disaster, you'll feel better. Similarly, don't worry about the things you can't change, don't expect yourself to perform perfectly at all times, remind yourself of the unimportance of trivial upsets, and don't assign the same priority (super-ultra-high) to everything on your to-do list.

- **Sleep.** As you learned in Chapter 3, a lack of sleep leaves your brain unable to regulate its emotions. If you're in this situation, you'll find yourself emotionally out of control and unable to rein in big reactions to little annoyances.

- **Seek social interaction.** Studies show that communal activities distract us and relieve stress. Sharing a laugh and nurturing a family member are particularly effective.

- **Assert yourself.** A key component of stress is a perceived lack of control. Without any way to respond, the brain's built-up energy is channeled into smoldering frustration. To avoid becoming a victim, speak your mind, make decisions, and be honest.

- **Perform good deeds.** Altruistic acts seem to help stamp out stress. The effect probably works because it boosts your sense of control, lightens your perception of the world, and enhances your self-image.

- **Know your stressors (and when to avoid them).** Sometimes a good debate is a lot of fun. Other times, it might send you careening over the edge, *Heart of Darkness*–style. And if you have any doubt about exactly what annoys you, ask your family members, who will have no trouble providing you with a full list of your mental buttons and the best ways to push them.

In Search of Happiness

The brain isn't interested in keeping its pleasure centers continually active. As you've seen, it uses flashes of pleasure and pain to keep you moving along the arc of your life. Unless there's something you need—for example, a warm coat on a wintry day or a jam-filled donut on an empty stomach— there's no way to get pleasure. That's because without need there's no desire, and without desire there's no gratification, and without gratification there's no hope of getting a zap in your brain's pleasure zone. Similarly, once your immediate needs are met, the pleasure dies off to make room for future goals.

What your brain craves, like virtually all of the systems in your body, is *homeostasis*—a perfectly even and unremarkable balance between you and your environment. When pleasure, fear, and other emotions disturb this equilibrium, the brain fights to get back on an even keel.

The Set Point Theory

In Chapter 2, you learned about the *set point* theory, which suggests the body uses every trick in the book to maintain its current weight. The depressing conclusion is that if your weight inches up over the years, you'll have a hard time fighting it back down.

The set point theory is just one example of homeostasis, and many researchers suggest happiness is another. To understand this theory, it's important to distinguish between *pleasure* (raw, physical feel-good feelings) and *happiness* (the more ambiguous state of contentment and optimism that we all generally strive for). Happiness is probably a secondary emotion that's generated in our deep-thinking cerebral cortex. In other words, pleasure is the biological drive that rewards our actions, while happiness is the subjective state we enter into when the conscious part of our brain reflects on our pleasure.

Here's the problem. According to the set point theory, our level of happiness is a basic personality trait. And much as your body fights to get back to its set point weight, your brain always drifts towards its set point of happiness. Some people are always cheery, no matter what apparent tragedies befall them. These are the people who don't mind being confined to a bed with kidney stones because it gives them a chance to catch up on their crossword puzzles. Other people study the dark lining in the happiest-seeming events. They worry about the tax implications of winning the lottery. Most people fall somewhere in between, and have brains that prefer a more moderate balance of moderate worry and mild satisfaction.

Even the huge life changes that you'd expect to create a long and lasting overhang of pleasure—say, inheriting an oil baron's fortune—don't change the recipe. You may have a few months of unrestrained excitement, and you may find that life gets easier, but the *joie de vivre* you feel from one minute to the next will soon settle back to your brain's natural level. Some studies suggest it takes as little as 3 months for a major change (say, moving from a dilapidated apartment to a palatial estate) to wear off. And as you'll see in Chapter 9, even the wild passion of a new relationship subsides to a calm and quiet bond after a couple of years.

The set point theory suggests you shouldn't be too hard on yourself when the sight of a cuddly kitten fails to stir a smile. More importantly, it suggests you shouldn't run yourself ragged in search of the things that you expect will make you happy (but probably won't have much of a lasting effect). In other words, if you're planning to quit your job, flee to Tahiti, and spend the rest of your life surfing the waves, all in a hedonistic hunt for happiness, don't bother. Your brain doesn't want it.

How to Accept Your Happiness Set Point

The situation isn't as grim as it seems. Once you understand that you may never be much happier than you are right now, you're ready to develop a broader perspective. After all, the brain is a complex place, and while you might not be able to stay in a state of endless cheeriness, you might be able to cultivate a deeper state of satisfaction. Here are some tips that show you how:

- **Redefine happiness.** More than a few thinkers have argued that happiness isn't endless joy, but something more along the lines of relaxed indifference. In fact, unremitting pleasure is largely a modern pursuit. Consider the word *nirvana*. Although it suggests banquet tables and orgies to most Westerners (okay, and a band, too), Buddhists use it to describe the complete peace of mind that one experiences when the brain is free from craving, anger, and pleasure.

- **Chase experience, not pleasure.** You can't guarantee happiness, but you can expand your sphere of experience. This is a great goal because it means you can cheerily look forward to giving birth, eating giant stick insects, and having a root canal (at least the first time). The experience-versus-pleasure equation is also what makes people value struggles like starting a business or raising children. These experiences often bring more worries than pleasure, but they also have the transformative ability to change who you are and the way you see the world.

- **Convince yourself you don't want to be happy.** It's not a complete stretch. After all, an eternally happy person would probably end up being lazy, incurious, unmotivated, and ready to take the entire world for granted. If your local chicken farmer was ridiculously happy, you couldn't get an omelet for breakfast. Unhappiness makes the world go round, because it fuels motivation, which leads to hard work, and occasionally generates progress. Finally, modern chemistry shows us that compounds that do bring unrelenting joy—say, heroin—have life-destroying power because they render the rest of our existence irrelevant.

- **Enjoy the freedom.** If you aren't obsessed with happiness, you don't need to try to buy it. This helps bypass a significant amount of disappointment, as happiness hunters inevitably forget that all luxuries have a time-limited effect, from granite countertops to cashmere sweaters. Otherwise, looking back in history would be looking back into a well of gradually increasing unhappiness, all caused by the depravation of essential modern conveniences like color television.

- **Remember that happiness is fleeting.** The word happiness is rooted in *hap*, the Middle English term for good luck (which also crops up in per*hap*s and *hap*penstance). Once you accept that you can't guarantee happiness, you can stop worrying about getting as much as possible and relax. This even-mindedness is the consolation prize for getting off the pleasure-reward treadmill.

- **Milk your happy experiences.** You only get so many moments of uninterrupted joy. However, you can stretch them over longer periods of your life using your brain's powers of expectation and memory. For example, if you're a gourmand who's planning to indulge in a $200 meal, mark the date in your calendar so you can look forward to the day. After it passes, squirrel the memory away in your brain so you can draw upon it when you're dining on local takeout.

Looking Forward and Looking Back

Most of the time, we don't consider the brain's remarkable ability to ratchet down pleasure and pain. For that reason, we're notoriously bad at predicting how happy a good event will make us feel (we overestimate) and how depressed a bad event will make us feel (we overestimate that too). This makes us pretty poor life planners. But now you know that your brain is pathologically unable to stay happy, and that knowledge brings some surprising power.

Most obviously, it suggests you shouldn't be afraid to take on new challenges. As you've already seen, the brain craves homeostasis. Whether you're in a loveless marriage or a dead-end job, your brain encourages you to sit quietly in the same spot. However, your brain also has a limitless capability to adjust to change. So if you've always dreamt of moving to Tobago and starting a mango farm for troubled youths, you'll be happy to know your brain is more than up for the task. (Just don't expect that it will transform your life into a nonstop pleasure party. The best you can hope for is that the experience will leave you enriched and full of tropical fruit.)

In other words, the power of adaptation is at work in every avenue of your life—so why not make it work for you? In modern life, we spend most of our time noticing how the brain steals pleasure away from us, but we're more reluctant to take advantage of its similar ability to neutralize fear, intimidation, and sorrow. This is a good short-term strategy, but a risky long-term approach to life, because it risks falling prey to another negative emotion: *regret*. Studies show that people are far more likely to regret things they didn't do than those they did. So if you make the move to the mango plantation in Tobago, your brain will happily rationalize it as a worthwhile step in your life. If you don't, there's a whole lot less for your brain to grip onto, and you'll be left always wondering about what could have been.

Note To see some colorful examples of how people overestimate the effects of various decisions on their future happiness, check out the book *Stumbling on Happiness* (Knopf, 2006).

Responding to Good and Bad

If you're an optimist, you're sure to stay that way. If you're a pessimist, there's no way to change you. Both outlooks are a reflection of our personalities and a deeply ingrained way of seeing the world. However, there's a gap between our attitudes (be they positive or negative) and our emotions. This gap is filled with a soft, slippery substance called *explanation*.

Studies show that people who tend to be deep, dark, and depressed internalize problems and favor something called *negative explanatory style*. Here are the hallmarks:

- When dealing with disaster, they identify themselves as the cause.

- When dealing with good outcomes they reverse the logic, and attribute success to random chance or external factors.

- They assume that ill consequences are pervasive (affect everything) and permanent (last forever). They assume good news is limited in scope and certain not to last.

On the other hand, optimistic people prefer *positive explanatory style*, and believe exactly the opposite. They treat happy times as their own doing and see setbacks as bad luck.

The following chart sums up the difference:

	Positive Explanatory Style	**Negative Explanatory Style**
Getting a Promotion	I deserved it. This is the beginning of a new direction in my life.	I got lucky. There's no way I'll meet their expectations for long.
Missing a Promotion	They were looking for different skills. I'll get it the next time.	They saw through me. This is the beginning of the end.

The bottom line is this: both views are biased. And while you can't will yourself to be more chipper, you can unmask your automatic judgments for what they are—bad habits—and replace them with more balanced assessments. If you're a resolute pessimist, you just might identify a few blind spots and learn to deal with negative emotions more effectively. And if you're a happy-go-lucky optimist, you might find your weaknesses and identify patterns of unsuccessful behavior before they cause serious damage.

The Practical Side of Brain Science

The Roots of Depression

No discussion of happiness would be complete without discussing the insidious disorder that saps pleasure from every reward—*clinical depression*.

At first glance, depression seems like a perfect candidate for a brain-based explanation. After all, scientists can identify distinct differences in the brains of depressed people. Most notably, they have lower than usual levels of certain neurotransmitters, like *serotonin*. Furthermore, depression is usually treated by drugs that raise the level of neurotransmitters in the brain by preventing them from being reabsorbed.

However, this simple description glosses over many mysteries that neuroscientists can't explain. For example, there is usually a delay of weeks or months before antidepressants have their maximum mood-boosting effect, even though neurotransmitter levels rise within hours of taking the first pill. Furthermore, dosages that are enough to raise neurotransmitter levels to normal levels are too low to have any effect on depressed patients.

Scientists now think that antidepressants trigger broader changes throughout the brain. For example, neurons may ratchet down their sensitivity to serotonin when they find that it's floating around like candy. Or, serotonin may act as a neuromodulator (page 17) triggering brain-changing processes in different parts of the brain. (Some believe it spurs neuron growth in the hippocampus.) In fact, there may well be a range of overlapping effects that kick in when neurotransmitter levels rise.

In any case, if you find yourself falling into severe depression—characterized by profound, constant unhappiness, lack of interest in the outside world, and suicidal thoughts—get help from a medical professional. Depression can't be self treated. However, if you're prone to depression but currently on an even keel, there's a lot you can do to mitigate the risks of relapse. Studies show that exercise, proper sleep and diet, solid relationships, a positive explanatory style (see the previous section), and activities that boost feelings of self-worth and belonging all help fend off mood disorders.

7 Reason

In the world of logic, it's easy to fault the human brain. For most of the day, we walk around with our critical brains powered down. We buy exotic exercise equipment from late-night infomercials. We pass around emails that link lung cancer to chewing gum. We send checks to pleasant Nigerian gentlemen with odd banking problems. Studies that track down the victims of these hoaxes don't just find bewildered seniors and lonely housewives—they also turn up lawyers, investment bankers, teachers, and other people who are in the business of thinking straight.

Sadly, the brain's shoddy thinking is more than a bad habit—it's an instinctive and automatic way of perceiving the world. When we hear a discussion, we filter out everything but the arguments we recognize and the ideas we like. Facts seep out of our brains like warm jello. We dive into health fads, fashion trends, new-fangled hobbies, political movements, and every sort of cobbled-together superstition that passes our way, all on the very thinnest of grounds. And when asked to explain our behavior, we look deep into our hearts and make something up. Quite simply, humans are masters of irrational behavior.

In this chapter, you'll learn why we often fall for sloppy thinking and fuzzy arguments. You'll see how quick assumptions, generalizations, and prejudices aren't just bad habits, they're also part of a critical set of life skills that helped our remote ancestors avoid ending up as another animal's dinner. Along the way, you'll uncover many of the worst reasoning mistakes that our brains make, and you'll learn to avoid them, compensate for them, and possibly use them to your advantage. Finally, you'll consider techniques for overdriving your brain with creative thinking.

The Thinking Brain

So far, this book has taken you to the near and distant corners of your brain. You've delved deep into its core to look at the *hypothalamus*, a critical piece of neural hardware that manages your appetite (Chapter 2) and controls your daily rhythms of sleep and of wakefulness (Chapter 3). You've also explored the middle ground, learning about the structures that encode long-term memories (the *hippocampus* in Chapter 5) and manage emotional drives like pleasure and fear (the pleasure circuit and the *amygdala* in Chapter 6). However, you've spent less time peering into the important topmost layer of the brain—the *cerebral cortex* that powers conscious thought. Oh, you've taken a look at how its unwritten rules shape your perception of sights, sounds, and other stimuli around you (Chapter 4), but you've yet to see how it deals with deductive logic, social dilemmas, and creative thinking.

Understanding the cerebral cortex is tricky, because important functions are scattered throughout its crinkly folds. Brain researchers can pick out dozens of specialized areas for tasks ranging from face recognition to speech comprehension. However, one area stands out for its role as a conscious control center, seat of high-level reasoning, and the home of your personality. It's the *prefrontal cortex* (PFC).

The Prefrontal Cortex

The prefrontal cortex is the portion of your brain that sits at the very front, just above your eyes and behind your forehead.

You already met your prefrontal cortex in Chapter 6, where you learned how it plays a key role in motivation. The prefrontal cortex also crams in a range of high-level mental processes. For this reason, it's often called the brain's *executive center* (presumably by people who actually believe executives do more than dine out on power lunches).

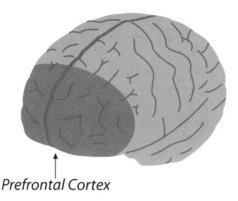

Prefrontal Cortex

Here are some of the tasks that the prefrontal cortex takes on:

- **Judgment.** The PFC supports the critical reasoning you'll learn about in this chapter. It helps you evaluate the good, the bad, and everything in between.
- **Choice.** The PFC lets you weigh different options, deal with conflicting thoughts, and make a decision.
- **Planning.** The PFC is keenly important for predicting the consequences of actions and setting long-term plans to reach specific goals.
- **Motivation.** The PFC helps you get things done. People with damage to the PFC often have severe trouble initiating new activities.
- **Social regulation.** The PFC helps you suppress inappropriate urges and lewd behavior, all for the greater good. Some studies have found that drug addicts, sociopaths, and criminals have weaker than usual connections between their PFCs and the rest of their brains.
- **Humor.** The PFC plays a role in our appreciation of fine comedy. People with damage to some parts of the PFC are more likely to appreciate slapstick humor, but have trouble understanding *double entendres*, puns, and subtler jokes.

 The PFC is one of the slowest maturing parts of our brain. Chapter 10 (page 232) explains that it's probably not fully mature in the average adolescent.

The prefrontal cortex is an exceedingly complex area of the brain. Neuro-logists have noticed that damage to the prefrontal cortex can cause a wider range of symptoms than damage to any other area of the brain. The prefrontal cortex is also a distinctly human specialty. Compared to other animals, our prefrontal cortex has been enlarged to vast proportions. (It's thought that over the past few million years, our brains tripled in size while our PFC grew a staggering six times bigger.)

In this chapter, you'll consider the role of your prefrontal cortex in reason-ing and social behavior. Half a century ago, the prefrontal cortex interested scientists for a different reason—they thought hacking it up could be a shortcut for dealing with adolescent insurrection, persistent moodiness, and overly outgoing wives (to name just a few well-documented cases). Indeed, it worked, in much the same way that amputating a foot cures bunions. The book *My Lobotomy* (Crown, 2007) by Howard Dully is a partic-ularly harrowing first-person account of the procedure and its aftermath.

 The story of Phineas Gage is a favorite example of PFC damage among neurology students. Phineas lived as a responsible and likable railway worker until a tamping rod flew completely through his head in a freak accident in 1848. Miraculously, though the rod entered under his chin and shot out the top of his head, it spared key brain regions and left him alive and able to function normally. However, Phineas was never quite the same. Without the executive control of the PFC, he become irresponsible, impatient, quick-tempered, and profane.

Now that you know where your deep thinking takes place, you're ready to face up to its flaws. First up: the foibles of common sense.

Common Sense

No one knows exactly what benefit early humans got out of their com-paratively enormous brains. Good theories suggest various possibilities—perhaps our pumped-up brains made us better foragers, hunters, cooperators, or romantic partners. However, it's clear that the brain first evolved for sur-vival and reproduction and has been thoroughly co-opted by the modern world, where it's used for distinctly non-life-or-death activities like chess, computer games, and existential Swedish movies.

This is important, because the human brain's way of reasoning is shaped by the needs of its ancient environment, and its occasional failures in the modern world are a legacy of that design. Thousands of years ago, every decision a human made had to be quick and was based on partial facts and second-hand information. So it's no wonder that we developed the perfect tool for making snap judgments with partial facts and second-hand information—namely, *common sense*.

The brain is an expert in common sense, which is the set of knowledge that everybody knows to be true because nobody wants to think about it anymore. Common sense has a pleasant face and a nasty underbelly. The good side is its blistering speed. It takes fractions of a second to conclude that you do want to pick up that $20 bill lying on the sidewalk, but shouldn't walk under a suspended piano to get it. The downside is its paunchy logic. In complex situations, common sense is all too often reduced to quick-thinking stupidity.

To get a handle on the problem, it's worth looking at some of the most common logical mistakes that your brain makes. In the following sections, you'll learn about the most common human biases—reasoning mistakes that we make automatically, instinctively, and constantly.

> **Note** The built-in biases of your brain aren't exactly mistakes. It's more polite to describe them as reasoning *shortcuts*. These simplifications let your brain respond quickly and decisively, which is essential in some situations but embarrassingly off-the-mark in others.

Anchoring

Your brain doesn't like to waffle. Rather than mulling a situation over, people prefer to make quick, provisional decisions, and then tweak these decisions with minor adjustments.

This can lead to a problem known as *anchoring*, where your brain fixates on one detail when assessing a situation, rather than considering the whole picture. For example, when contemplating a new home your brain may latch onto a few compelling features, such as a remodeled kitchen or hot location. Your brain will then gloss over other potentially negative details, such as its high price, its old age, and the teenage tenants playing *Guitar Hero* next door.

Tip To overcome anchoring, you need to train your brain to hold off its decision making and entertain possibilities that may initially seem like dead-ends or bad ideas. Page 163 explains the thinking techniques that can help you out.

Conservatism

Conservatism describes how people tend to hold onto their opinions even in the face of new, contradictory information. (Insert your own joke here about the Republican Party; this section, however, has nothing to do with political movements.) In the pre-historic world, conservatism was a sensible strategy. New information was uncertain and untested, so the brain placed a premium on long-held beliefs. In modern life, conservatism makes us more likely to ignore new facts and hold onto old habits. For example, although it's well established that nitrates are linked to cancer, it's easier to dismiss the science than change entrenched eating habits that favor bacon-wrapped hotdogs.

A certain degree of conservatism is healthy. For example, consider how we react to scientific research, which loses quite a bit of contextual information as it trickles down through popular media. This contextual information—including details such as the size of the study, the way it was administered, and how it agrees with other research—is what experts use to separate repeatable, agreed-upon conclusions from promising new ideas and the wild rantings of grant-starved researchers. Without this broader picture, it's best to hold off a bit before you accept the conjectures of new research (although you'll get some guidelines for evaluating new ideas on page 165). After all, if you tossed out your avocados, oil, and butter when the low-fat diet craze hit, shed your steaks and chicken when vegetarianism was in, and then ditched fruit, beans, and bagels when Atkins ruled the diet world, you'd be left with a pretty bare pantry.

However, watch out for the cardinal sin of conservatism—the tendency for people to place more weight on information confirming what they expect than information that contradicts it. If you think smoking strengthens lungs and the U.S. moon landing was a poorly staged walkabout on a beach, you probably rely on an unhealthy dose of conservatism to maintain your wonky beliefs.

The Mere Exposure Effect

People prefer things that they're familiar with. Advertisers rely on this *mere exposure effect* to underpin irritating ad campaigns that actually do make you more likely to buy their products. Once again, this bias may have its roots in our deep, dark evolutionary past. In pre-historic times, anything new was a potential source of harm. But if something lingered around for a while without killing anyone, it was probably safe.

Grouping

In its quest to understand the world around you, your brain struggles to categorize *everything*. After all, if you understand that pigs are pink, rotund, and tasty, you don't need to remember the individual details about every single one you meet. Instead, you can head straight for the frying pan.

In other words, grouping is one of the tools that your brain uses to reduce huge quantities of information into practical rules that you can put to use in everyday life. Humans are successful in many walks of life because they're excellent groupers.

Unfortunately, we often *over-group*, and once items are placed into categories we no longer perceive them in the same way. Our brains automatically emphasize the differences between groups and minimize the differences between members of the same group—even if we have to stretch logic to do so. This is true even if the groups are completely arbitrary. For example, studies that split people into made-up groups (for example, red-shirt-wearers and blue-shirt-wearers) find that participants exaggerate the differences between the groups and minimize the differences inside the groups just as readily as they separate pork from beef.

 Note There are some good reasons for bad prejudices. In pre-historic life (and arguably in competitive modern environments such as business and sport), clinging to your groups with solidarity and reacting suspiciously to outsiders is a good survival strategy.

Unfortunately, the brain's grouping bias frequently spills over into distinctions that are easy to make but have little significance. For example, when you meet new people at a dinner party your brain will automatically categorize them based on race, profession, gender, age, geographic home, level of attractiveness, and income bracket. You'll then be tempted to apply assumptions based on these categories, particularly if they aren't the same categories that you fall into.

The most obvious example of human grouping run amok is, of course, *racism*, the tendency to generalize about other people based largely on the level of melanin in their skin. One reason that racism is difficult to combat is that people look so obviously different from one another. A thinking brain automatically creates categories based on physical characteristics like skin color and facial features. And as you've learned, once your brain makes these distinctions it can't help but use them to attach assumptions.

Late Night Deep Thoughts

Do Races Really Exist?

We all know that it's impossible to predict the personality of a specific individual based on details like skin color. But what about broader studies that attempt to dig up statistical differences in different racial groups. Do these make sense?

From a scientific standpoint, they probably don't—at least not in the tidy way we'd expect. The first problem is that it's nearly impossible to tease apart the influence of culture and genetics. For example, Thai people are more likely to enjoy Thai cuisine, and visitors from India are more likely to practice Hinduism, but neither association tells us anything about race.

To really zero in on racial differences, you need to dig into the science of genetics. However, when we lay one race's DNA against another's the problems really start to stack up. For example, the genetic differences between different groups of Africans are far greater than those between so-called white and black people. In other words, even if it's possible to divide people into different genetic populations, our attempt to do it with the groups we call races isn't on the mark. Furthermore, the human race as a whole has far less variability than many other species, including dogs and chimpanzees.

The bottom line is that humans differ in many ways. The concept of race captures a sliver of that diversity, but it also distorts it, emphasizing differences that are trivial and implying similarities that don't exist. Lastly, race is also a social construct. When settlers landed in America, they viewed themselves as belonging to several very different ethnic groups, and only gradually banded together under the newly invented category of whiteness. The same effect persists today—the need to belong to a group often overrides any quibbles about how that group is defined.

Moral Calculus

Common sense really takes a counter intuitive turn when we attempt to make practical decisions about moral issues. The psychologist Jonathan Haidt has a great deal of fun testing people with brain-bending moral problems like the ones shown here.

Before continuing, answer the question "is this morally right?" for each scenario, and then think of a quick one or two sentence explanation that backs up your reasoning.

A woman is cleaning out her closet, and she finds her old American flag. She doesn't want the flag anymore, so she cuts it up into pieces and uses the rags to clean her bathroom.

Julie is traveling in France on summer vacation from college with her brother Mark. One night they decide that it would be interesting and fun if they tried making love. Julie was already taking birth-control pills, but Mark uses a condom, too, just to be safe. They both enjoy the experience but decide not to do it again. They keep the night as a special secret, which makes them feel closer to each other.

A man goes to the supermarket once a week and buys a dead chicken. But before cooking the chicken, he has sexual intercourse with it. Then he thoroughly cooks it and eats it.

Most people feel that these scenarios are morally wrong. When asked to explain why, they trot out some reasons that sound good—for example, incest can cause birth defects and eating a post-coital chicken is unsanitary. Of course, a more detailed look at the scenarios shows that they've been specifically constructed to outwit these objections. Julie and Mark are careful to ensure there's no possibility of pregnancy. A thoroughly cooked chicken with a little, um, extra protein doesn't pose a health risk. But if you're one of the many that's repulsed by these ideas, these logical arguments won't make you feel any better. When confronted with these counterarguments, study participants didn't change their minds—instead, they simply looked for different reasons to support their conclusions.

Essentially, these examples show how the brain prefers rationalizing to reasoning. Rather than fully evaluate a situation, it prefers to leap to an instinctive conclusion and then think out arguments to defend it. In the case of the moral-testing examples, the scenarios activate deeply ingrained reactions that favor social norms. And social norms aren't just fluff—they underpin humanity's great transition from small wandering groups to complex societies. So it's no surprise that solid social instincts are a part of the brain's automatic programming.

 Note To try out more moral-testing dilemmas and be part of future studies, visit *www.yourmorals.org*.

Incidentally, brain scans show that pro-social responses involve a small region of the brain called the **ventromedial prefrontal cortex**, which sits inside the PFC.

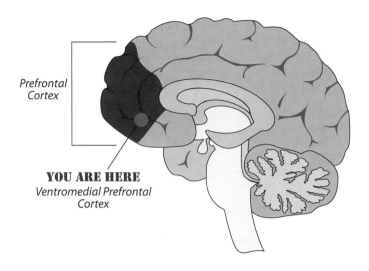

Prefrontal
Cortex

YOU ARE HERE
Ventromedial Prefrontal
Cortex

If the ventromedial prefrontal cortex is damaged, people are more likely to accept behavior that threatens social norms but doesn't cause real harm. They also have an easier time making coldly calculating decisions that **do** cause harm, but maximize the welfare of a group of people. (Some of the moral choices that brain damaged people are more likely to endorse include halting a runaway train that's headed toward a group of people by throwing a fat man in its path, murdering an innocent person to harvest his organs in order to save many more sick people, and smothering an infant before his crying can alert enemy soldiers to your family's hiding place.) These studies suggest that the ventromedial prefrontal cortex is the part of the brain that overrides mere reason with pro-social emotions like compassion, embarrassment, and guilt.

Statistical Blunders

Statistics is about as far as you can get from the brain's common sense thinking. As you probably know, **statistics** is a set of mathematical techniques that draws certain types of conclusions from huge quantities of information. These days, we use statistics to inform everything from what shows we put on television to what medicines we put in our bodies.

Unfortunately, the human brain is embarrassingly bad at thinking statistically. Your brain is far happier relying on a hodgepodge of hunches, best guesses, and personal experience than it is analyzing numbers and trends. As a result, we're often unable to take full advantage of the best information we have about the world around us.

The brain's preference for instinct over statistics makes perfect sense. For millions of years, humans had no need to think statistically because there were no statistics. Furthermore, if one of our distant ancestors had taken a day off to invent statistics, it would have been profoundly useless, simply because there would have been no way to gather the huge amounts of information needed to make statistical conclusions. In other words, humans are experts in making calculated assumptions based on limited information because we need to be. It's only in the last few hundred years that we learned how to nose into millions of other people's lives to help make decisions about our own.

Small Samples

Today is a special day in Ted's life. After a protracted struggle with a nasty smoking addiction, he's finally decided to ditch the habit. Not one to let the moment pass, he immediately sets off to his local pharmacy to buy a nicotine patch and is promptly run over by an oil tanker.

The question is this: was it a good idea for Ted to quit smoking? Clearly it was a bad move—in fact, it's no exaggeration to state that Ted's decision to kick the habit resulted in his untimely demise. But this odd turn of events should bring little cheer to Joe Camel, as it doesn't make the slightest difference in the overall statistical conclusion that, all things considered, smoking is a surefire way to end up sicker and die sooner. Ted's life is just one data point in a huge mass of information.

As statisticians know, individual examples are almost meaningless when attempting to unearth cause and effect relationships with statistics. You can start every day with a two-Red-Bull-breakfast and end it with a half-liter vodka nightcap, and still live past 100. Or, you can adopt the strictest vegan diet and die of colorectal cancer while your diabetic Twinkie-chomping family members look on pityingly. Life is odd that way.

That doesn't mean, of course, that you should down immense quantities of caffeine, alcohol, and synthetic sponge cakes. Doing so only increases the chance you'll suffer various health problems—and all things being equal, it's more fun to gamble when the odds are in your favor. However, your brain doesn't always understand the difference. Left unchecked, it prefers to reason from the here and now. Only statistics can help you understand the subtler patterns that emerge over long periods of time.

Selection Bias

One of the most notable mistakes we make when sizing up information is to pick and choose what facts we consider and which ones we ignore. This bias is unintentional—it just so happens that unusual events stick in our memories more effectively than ordinary ones. This is called the *selection bias*, because we *select* what facts to consider when we draw sweeping conclusions.

For example, imagine you have a vivid dream about the death of a dear uncle. A few days later, he dies. It's hard to avoid the feeling that your dream picked up on some mysterious current of paranormal energy, and predicted the future. A paranormal-doubting skeptic might try to tell you that your dream was the result of subconscious parts of your brain that knew your uncle was likely to die because he was an elderly man, a cancer patient, or a tightrope-walker. But in fact, none of these arguments are needed to explain what happened. Statistics can clear up the mystery much more effectively:

- **Eventually the unlikely has to happen.** Raw probability tells us that unless there's some supernatural power at work, some random dreams will eventually turn out to be true. Right now, billions of people on planet Earth are dreaming, and hundreds of thousands of them are dreaming about the death of a loved one. Just by chance, some of these relatives will pop off in the next few days. In the broad statistical picture, this isn't particularly remarkable. But in an individual person's life, the effect is much more dramatic. In fact, dying and dreaming are both so commonplace that it's inevitable that a few unlucky people will have the predictive dream experience several times in their lives.

- **We aren't including all the facts.** The selection bias means that we'll remember a predictive dream for far longer than another dream that doesn't turn into reality. In other words, if we statistically examined *all* our vivid dreams to determine how often they predict an unexpected event, we'd find that our success rate is pretty poor. But because a predictive dream is such a powerful experience, we'll remember it long after we've discounted the time we dreamt about buying bagels in the nude.

 Note The selection bias is the reason why meeting a distant acquaintance in a shopping mall seems like an unbelievable coincidence. In this situation, we can't take other examples into account because we don't know about them. For example, we have no knowledge of all the times that we might have been in the same place as a long-lost friend, former lover, or sworn enemy, but missed a meeting by mere seconds.

The selection bias is an insidious part of everyone's day-to-day reasoning. Here are a few more examples. Can you spot the skewed sample?

❶ *The only time I spill something on myself is when I wear white.*

❷ *Everyone who goes to this hospital ends up sick, and most of them die.*

❸ *I wouldn't go to America because of all the gun crime; we see it on the news all the time.*

❹ *A recent study found 85% of people who overcame cancer prayed daily and believe God healed them.*

❺ *There are more black men in prison than in college.*

Here are the answers:

❶ You probably don't remember the relatively harmless times you spilt something on your black jeans.

❷ Sick people are more likely to go to the hospital in the first place.

❸ The news media selects shocking events. The fact that it focuses on gun crime doesn't imply that gun crime is common. It simply implies that no one wants to watch news about, say, a 70-year-old who makes it home from her afternoon bridge party without getting mugged. (That said, the rate of gun-related deaths in the U.S. *is* statistically higher than most other countries due to its high rate of gun possession, but the chances of you encountering a bullet on your visit is close to infinitesimal.)

❹ But what about the people who died? Perhaps 95% of them prayed twice a day. This mistake, known as the survivor bias, is used to defend an endless parade of miracle cures.

❺ This statistic deftly implies that more black men will go to prison than to college *in their lifetimes*. But on closer examination, it actually says something quite different. The problem is that this statistic compares different samples that have nothing in common. The group of black men in college is drawn from a relatively small group of college-aged black men. The group of black men in prison draws from the much larger group of *all* black men. Furthermore, college lasts an average of 3 to 4 years, while a prison sentence can last 10 years or more. So this statistic omits many college educated, law abiding black men who aren't currently in school.

 Note Selection biases aren't always accidental. Sometimes they're deliberately used to create statistics that sound compelling but have little meaning. For help training yourself to spot the misuse and abuse of statistics, read a book like *How to Lie with Statistics* (Norton, 1993).

Regression to the Mean

Imagine you stumble across a Web site advertising a miraculous magnetic bracelet that heals minor ailments. Seeing as you're suffering with a cold, you fill out the order form. As you wait for the order to arrive, you begin to feel better, and when the bracelet appears four days later there's no sign of your cold. Clearly, the magnetic bracelet cured you—in fact, its mysterious influence healed you before you even slapped it on your wrist, right?

The magnetic bracelet example might not seem terribly convincing to a clear-thinking brain. But the brain craves patterns, and imagining relationships that don't exist—for example, a magnetic bracelet that cures the common cold—is one of its favorite pastimes.

A more devious example of this tendency is found in a phenomenon known as the *regression to the mean*. To understand how it works, imagine the following list describes your test scores in a neurology class:

Quiz 1:   78%
Quiz 2:   74%
Quiz 3:   59%
Quiz 4:   72%
Quiz 5:   70%
Quiz 6:   85%
Quiz 7:   77%
Quiz 8:   74%
Quiz 9:   77%
Quiz 10: 72%

For all 10 quizzes, your average score (or *mean*) is a respectable 73.8%. But look at what happened with the third quiz. Here, you suffered a cringe-inducing 59%. (Don't worry; it happens.) On the next quiz, you scored your more typical 72%. So why did you improve? Statistically speaking, every sequence of numbers includes a few outliers. So after a bit of bad luck, your test scores returned to normal, returning closer to your 73.8%. Technically, they *regressed to the mean*. The same thing happened after your hotshot 85% score on the sixth quiz.

Here's the problem. The pattern-matching circuitry in the brain has a lot of other ways to try and explain your fluctuating performance. It might correctly pin your poor showing in the third quiz on a late-night bongo party the night before. It might also invent more imaginative explanations for your more typical fourth test performance. For example, maybe you studied with a braniac friend, fasted for 24 hours, or donned a magnetic bracelet. From a statistical point of view, the rise after the third quiz and fall after the sixth quiz are just ordinary fluctuations. From the brain's point of view, they're distinct events that must have some clearly identifiable cause.

Regression to the mean has one particularly destructive effect. It encourages people to react more dramatically to poor events than positive ones. For example, imagine your neurology instructor pulls you aside for a pep talk after the third quiz. He'll be rewarded by seeing you raise your game the next time around. But if he praises you after your sixth quiz, statistics won't grant him the same benefit—instead, he'll watch you plunge back to average. After this happens a few dozen times, he might decide to keep his thoughts to himself when dealing with high performers, and spend more time talking to the laggards.

 Note In many cases, the regression to the mean effect encourages people to be strict punishers and faint praisers, despite the fact that studies show praise has a more positive effect on learning.

Probability

Probability theory, which analyzes the likelihood of certain events, is one of the best examples of our brain's trouble with statistical thinking.

To see the problem, try out the famous Monty Hall problem, which presents a probability challenge loosely based on an old gameshow.

Suppose you're on a game show, and you're given the choice of three doors: Behind one door is a car. Behind the other two doors are ill-tempered goats. You pick a door (which we'll call door 1) but leave it closed.

The cunning host knows what's behind all the doors. After your choice, he opens another door (which we'll call number 2) to reveal a goat. (This is a standard routine that the host follows in every game.)

The host then asks you if you want to switch your initial choice from door 1 to door 3.

Is it to your advantage to switch?

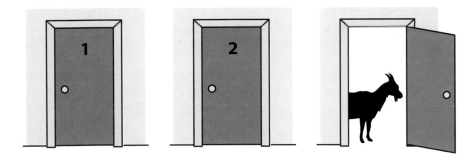

This tricky problem is counter-intuitive, and has been the downfall of many math professors. (In fact, hundreds of them wrote in with faulty logic to "correct" the solution that was given in *Parade* magazine in 1990.)

The most common (and thoroughly incorrect) answer is that it makes no difference, as each remaining door has a 50% chance of leading to the car. The right answer is that switching is by far the best bet—it doubles your chances from 1/3 to 2/3.

Still stumped? The often missed ingredient is the fact that the host is actually helping you out by opening a failing door. (The host never opens the door that leads to the car, because that would be anticlimactic.) The easiest way to size up the situation is to consider every possible way this challenge could unfold, using a probability tree like the one shown here.

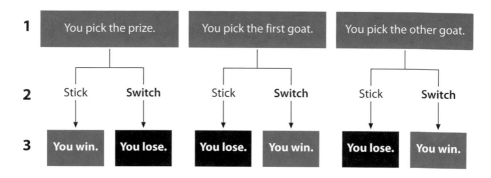

Here's how it breaks down:

❶ When you initially pick the door, there are three equally likely possibilities.

❷ Then, you have a decision to make.

❸ Switching wins two out of three times. In fact, switching only loses if your first guess was on the money, and the chance of that is just one in three.

To make the problem even more compelling, assume there are 100 doors and the host opens 98 goats after you make your initial selection. In this version of the problem, you've got a virtual lock on the prize if you switch. And if you're still in doubt, try a guess-where-the-coin-is game with three cups and someone else's money. Nothing brings probability into focus like the ability to beat someone else's pants off.

Random Events and Games of Chance

The Monty Hall problem demonstrates how people have trouble converting statistical information into the right decisions. These problems are particularly frequent when people need to deal with random events. One example is the so-called *gambler's fallacy*, the assumption that the longer something doesn't happen, the more likely it becomes. For example, a gambler watching a game of roulette might become convinced that number 14 is due to turn up soon, because it hasn't been hit for several rounds.

 Note If gambler's fallacy were correct, you could easily cheat at heads and tails by privately tossing a coin. Once you produced a sequence of heads, you could find a friend and make a rich bet that the next toss will turn up tails, secure in the knowledge that your coin is biased in favor of this result thanks to your pre-tossing.

Gambler's fallacy is another example of the pattern matching circuitry in your brain working overtime, trying to make sense of random information with pseudo explanations. (For example, "Joe's on a winning streak," or "The red ball is *due*," or "My luck's bound to change soon.") The remarkable fact is that even though gamblers spend hours playing probability games, their experience can't teach them the laws of probability. Probability is simply too counter-intuitive—and the brain is more than happy to seize on red herrings, distractions, and selective memory to reinforce its faulty thinking.

 Note Although you've now reached the end of this book's exploration into human biases, you've by no means exhausted the catalog of brain bungles. For a massive list of mistakes, check out *http://en.wikipedia.org/wiki/List_of_cognitive_biases*, which describes errors in decision making, probability, social interactions, and memory.

Critical Thinking

So far, your exploration into reasoning and the brain has been more than a little depressing. With its impressive catalog of logical errors, leaky assumptions, and glaring omissions, it's a wonder you're able to make toast without a textbook at hand.

Fortunately, you can train your brain to behave more rationally. The following sections present some of the best practices of *critical thinking*, which provides a disciplined way to challenge common sense.

Accepting Uncertainty

As you've already seen, your brain has a deep hunger for certainty. It's most comfortable with concrete, actionable information, and it barely tolerates ambiguity. Rather than use logic to openly investigate an issue, your brain prefers to latch onto a conclusion instinctively and use logic after the fact to defend it. The colorful writer Edward de Bono describes it this way: "The natural tendency of thinking is to support a view arrived at by other means."

To battle this tendency, you need to master the art of suspending judgment. The longer the interval between the time a question is posed and the time your brain locks into an answer, the more objective you'll be. Once your brain forms an expectation, that expectation acts like a magnet, pulling all your thinking and reasoning in one direction.

In some cases, you'll need to accept the ambiguity of having no clear answer at all. To do so, you need to fight against your brain's instincts, which favor bad explanations to no explanation. This tendency underlies everything from wrongful convictions to wacky superstitions.

Overcoming Bias

The most important step in critical thinking isn't applying logic in a particular way, but establishing the right foundation. You need to create an environment that leaves space for your brain to think.

As you've just seen, part of the trick is preventing your brain from making any preliminary conclusions. Another key point is to recognize your own fallibility. Although you can't erase your personal biases, you can keep them in check by adopting a mindset of intellectual humility. Here are some points to help:

- Accept that everyone has subconscious biases. Question anyone's automatic judgments.

- Remind yourself of strong beliefs that you once held but now reject.

- Admit that there are blind spots in your own perception of reality.

Baloney Detection

The world is drowning in shoddy ideas and half-baked arguments. The most important application of critical thinking is being able to separate the meat from the baloney.

Although there's no litmus test to evaluate ideas and opinions, a healthy dose of skepticism is essential. Many thinkers have weighed in to suggest techniques that you can use to recognize the slightly funky smell of an argument that's past its due date. (The astronomer and science popularizer Carl Sagan described his Baloney Detection Kit in a book called *The Demon Haunted World* [Random House, 1996].) So the next time you hear about a new and contentious theory, keep these questions in mind:

- **Can the theory be disproven?** It's easy to come up with ideas that can't be contradicted. Here's one: "Elvis is alive, but there's a conspiracy to hide the truth. Every piece of evidence that suggests he died is an elaborately constructed deception by an alien race." Theories that can't be disproven might make us feel better, but they just complicate life.

- **How reliable are the facts?** Ideally, facts should be confirmed independently. The gold standard in medical research is the *double blind* experiment, where neither the person being examined nor the doctor doing the examining knows which treatment is being used on which patient.

- **What do other studies say?** Does this new idea appear to contradict some solid and longstanding information? If so, there should be an explanation about what went wrong before, and it should be possible for other researchers to assess the new information and duplicate the results on their own.

- **Are there alternate explanations?** The world is full of tantalizing relationships that have murky causes. For example, students who study longer hours get better test marks. This suggests that studying raises grades, but it might just as easily reflect the fact that smart students tend to spend more time studying, students who have trouble give up after the first hour of review, or students tend to overestimate the work they put in when they excel and downplay their effort when they go down in flames.

A key touchstone in critical thinking is that ***correlation doesn't imply causation***—in other words, things may happen together even when one doesn't cause the other. For example, the 1990s saw increases in the rate of church attendance and drug use. So does prayer cause pill popping? Obviously, there are many more possibilities—perhaps substance abusers turned to God, perhaps both increases were the result of a third factor (such as social unrest), or maybe it's nothing more than a coincidental overlap of two separate phenomena. And although this example seems insultingly obvious, it's no different to similar connections drawn between vaccination and autism, or television and violent crime.

Winning Arguments with Logical Fallacies

Most discussion doesn't involve critical thinking. Instead, it revolves around a few rhetorical strategies that are designed to satisfy the brain's instincts and play to its logical weaknesses. Because a lazy brain is a happy brain, these tricks usually work better than a deep and nuanced debate.

The following list summarizes the best you'll find in any politician's toolkit of dirty logic tricks. It's between you and your own brain whether to use this list to expose fraudsters or persuade the unwitting.

Use value-laden words

If you construct your argument using loaded words that convey certain values, others will make snap decisions in your favor. For example, the same side in a civil war gets more support when described as ***freedom fighters*** than ***rebels*** or ***insurrectionists***. Similarly, if you don't like the science curriculum for your local college, you can point out that it's set ***undemocratically*** by a group of ***elites***. If you do like it, you can praise the fact that it's determined by ***respected experts*** rather than the ***ignorant masses***.

Most loaded words have tiny, knee-jerk arguments embedded inside them. Other loaded words that you might want to use to anchor your arguments include ***traditional values***, ***activist judges***, and ***tax relief***.

 Most political speeches are simply strings of value laden words, draped around the void where an idea should be.

Attack the arguer instead of the argument

Some examples include discounting a priest's views on abortion because he's a priest, dismissing a friend's treatise against animal cruelty because he's enjoying a smoked brisket, and disregarding a colleague's stance on capital punishment because he changed it twice last week. People may have vested interests, they may be hypocritical and easily swayed, but all these details simply distract from the real question of whether an argument is logically sound.

 Note This fallacy is also known as an ***ad hominem*** attack (an "argument against the man").

Argue from authority

This is the reverse of the attack-the-arguer fallacy. Here, an argument is defended on the basis of tradition, the majority opinion, so-called experts or important people who share the view, and so on. But logic isn't a popularity contest, and the validity of an argument doesn't depend on the people who are involved in promoting or opposing it.

A related trick is to condescendingly point out the differences between you and your debater. Explain to your opponents that they would understand if they were rich, if they shared your life-altering experience, if they were older, or if they had your hard-scrabble upbringing.

Distort the opposing view

It's easier to win an argument against a warped and exaggerated version of your opponent's view than the real deal. For example, claim the opponent's vote against war funding is a betrayal of the troops. Describe evolution as the idea of a dog giving birth to a cat.

 Note This is also known as the ***straw man*** fallacy, because it aims to create a version of the opposing view that can be knocked down as easily as a person made of straw.

Shift the goalposts

There are many different ways to argue for or against an issue. Every argument has its weaknesses and its strong points. An underhanded debater can jump from one line of argument to the next, getting the best of all of them, but following none of them all the way through and exposing it to counter arguments. For example, if you want to take down a local environmentalist, you might try a line of reasoning like this:

> "Global warming isn't happening. And even if it is happening, it's not that bad. And even if it is that bad, it's too expensive to fix it."

This argument's strength, such as it is, lies in its ability to throw down a cloud of ambiguity that skilled speakers can use to their advantage. If any one of these three overlapping points is called into question, one of the other themes can be pulled into the picture to reframe the debate and disorient the opponent.

Present a false choice

Don't allow ambiguity. Disable fine distinctions by forcing black and white logic. A famous modern example is the statement "Either you're with us or you're with the terrorists," which suggests that the entire world can be separated into two neat categories, one with trusted allies and the other with nefarious enemies. False choices are usually at work when someone uses the words *no alternative* or *slippery slope*.

False choices also help slick debaters use the flaws in opposing arguments as though they were the strengths of *their* arguments. This is a standard trick in political debates, and if you pull it off quickly you'll never get caught. For example, imagine someone argues this:

> "Gay couples can't have children, therefore they can never have true marriage."

Sensing a weakness in the argument, you can simultaneously refute it and use the false choice fallacy to take the upper hand:

> "As a society we don't prevent marriage between menopausal women and impotent men. Obviously, marriage doesn't need the possibility of procreation. Therefore, gays deserve to have equal marriage rights."

In this example, the first speaker's point has been successfully refuted. However, it's not correct to conclude (as the second speaker does) that the debate has been decided in favor of same-sex marriage side. The truth is the issue remains undecided until someone can put forward an argument that can't be shot down.

Criticize the consequences of a belief

Put pressure on your opponent by claiming their beliefs will doom us all. This fallacy was famously used by Pascal to point out that those who don't believe in God are likely to be left out of the party if He does exist. Pointing out the bad implications of good arguments is a particularly effective way to fight bad news, such as the claim that a war is going poorly ("We can't afford to lose") or that the Earth is heating up ("Our economy can't afford to change").

Use circular logic

Everyone's favorite logical trick is to game the system by assuming what you need to prove. For example, "The story of Adam and Eve has to be true because God would not deceive us." Or, from the opposite side of the religious spectrum, "Miracles can't happen because they would defy the laws of nature."

Problem Solving

Critical thinking is the best tool when you need to challenge a daft idea with withering logic. Critical thinking excels at testing, challenging, and dissecting minute details. But it isn't the right instrument if you need to create something new or get a fresh perspective on a perplexing problem.

Consider the following challenge. The game is hangman, and the category is films. The goal is to guess the movie title by filling in the four blanks underneath the gallows before exhausting all your guesses.

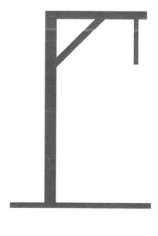

In a typical game of hangman, one body part is drawn in after each wrong guess. At its most challenging, only six guesses are allowed (one head, one torso, two arms, and two legs), after which the hangman is complete. However, you can give yourself 10 guesses. After each guess, consult the answer key shown below to see if you've chosen correctly. Try it now (and don't continue until you want to know the answer).

Answer Key

The Letter Grid lists all the letters of the alphabet, with a number after each one. If you wish to guess a letter, make note of this number and look it up in the Position Grid underneath to get a second number. This second number indicates either the position of the letter in the puzzle or 0, which means the letter doesn't appear in the answer. For example, in the Letter Grid the number for A is 1. In the Position Grid, position number 1 gives a second number of 0, which means there is no A in the puzzle. (That one's free.) If there were an A and it was the second letter, the Position Grid would give the number 2 instead of 0.

Letter Grid

A:1	B:4	C:1	D:8	E:5	F:9	G:4	H:8	I:5
J:8	K:5	L:9	M:4	N:1	O:5	P:8	Q:4	R:1
S:9	T:1	U:5	V:4	W:9	X:8	Y:4	Z:9	

Position Grid

1:0	2:4	3:3	4:0	5:0	6:2	7:1	8:0	9:0

Still stuck? The answer is Stanley Kubrick's landmark film *2001*.

Yes, the question wasn't quite fair. However, if you look back at the discussion of the game, there's no mention made of letters rather than numbers. This challenge demonstrates how invisible assumptions constrain your thinking. You aren't able to critically challenge these assumptions because they slip under the radar.

 Note Creative problem solving is often described as thinking "outside the box." However, it's more accurate to say that successful problem solving involves thinking *about* the box—in other words, finding the unwritten assumptions that limit you so you can break through them.

Ready to try again? Consider the following riddle:

> A young boy and his father were out playing football when they were caught at the bottom of a giant pileup. Both were injured and rushed to the hospital. They were wheeled into separate operating rooms and two doctors prepped to work on them, one for each patient. The doctor operating on the father got started right away, but the doctor assigned to the young boy stared at him in surprise. "I can't operate on him!" the doctor exclaimed to the staff. "That boy is my son!" How is this possible?

Here's the answer. As you read the story, you construct a mental picture of what's taking place. That mental picture includes the two doctors. The first image that springs to mind is that of two male surgeons, because your brain correctly calculates that this is by far the most likely case (after all, more than 90% of surgeons are men). Once established, this mental image blocks out other reasonable alternatives, such as the possibility that the doctor in question is a female.

The problem isn't that it's hard to imagine a female surgeon—had the story included this information, you wouldn't have stopped to think about it. But no matter how reasonable, your brain's automatic assumptions blind you to other possibilities. (You also saw this behavior on page 88, when interpreting a culturally ambiguous picture.)

This sort of puzzle is often called a *lateral thinking* puzzle, because it forces your brain to take creative tangents that divert from the step-by-step progression of ordinary logic. Once they've been revealed, lateral thinking puzzles seem embarrassingly obvious. Unfortunately, you can't solve them with raw critical thinking power alone. Instead, you need a few creative thinking tricks to free your mind.

 Note Creative thinking puzzles can be played in a group setting, where one person knows the puzzle and answers "yes" or "no" to questions posed by the other people. This form of the game, called a *situation puzzle*, combines creative thinking with the step-by-step Socratic questioning of critical thinking. To try some free situation puzzles, check out *www.kith.org/logos/things/sitpuz*. To practice some more lateral thinking puzzles, pick up one of the many puzzle books by Paul Sloane.

Creative Thinking Tools

Creative thinking tools are tricks that help you break out of the shackles of ordinary thinking. They short-circuit the highly efficient but somewhat single-minded patterns of automatic thought that rule your life.

Creative thinking tools are particularly useful when you need to solve lateral thinking problems (like the challenges described in the previous section) and when you're looking for a new way to attack an intractable problem. You can think of them as recipes designed to help you cook up new ideas. They aim to:

- **Stimulate your creative side.** If you generate enough new ideas, eventually one of them will make sense.

- **Distract you from your biases.** Games, role playing, and other tricks help you ease the focus away from yourself. That way, your instincts and opinions won't limit the alternatives you consider.

- **Reframe the problem.** Often, your brain traps a problem in a web of assumptions. By taking a radically different perspective, you can shake it free.

 Note Studies of improvising jazz musicians suggest that when you get creative, part of your prefrontal cortex shuts down. In other words, the executive centers of planning, judging, and control need to get out of the way when it's time for a creative jam session.

The Art of Provocation

The most creative problem solving begins when you challenge an existing assumption with a new idea. Unfortunately, it's all too easy to rule out an apparently absurd new direction before it gets off the ground. To check this habit, you need to master the art of provocation.

One technique invented by the creative thinker Edward de Bono is to use the word *po* to signal that a new idea is a *provocation* that can't be judged, but has to be used as a springboard for new ideas. For example, imagine you're trying to use creative thinking to figure out how to attract new customers to a faltering restaurant. You might use po like this:

Po: Let's admit we suck.

Po: Let's force people to come.

Po: Let's stop selling food.

These ideas are obviously illogical. But when you're forced to grapple with them, you might arrive at the following new ideas:

Po: Let's admit we suck.

> We could call the restaurant The Dive Bar. The ironic insult might attract more of the hip young professionals in the neighborhood.

Po: Let's force people to come.

> Why not sell food to prisons and elementary schools? Then we have a captive audience, and they don't have any other options.

Po: Let's stop selling food.

> Maybe we'd be more successful as a retro dance club with a liberal bring-your-own-chow policy.

You won't arrive at these new ideas immediately, but you can see how forcing yourself to consider the initial premise leads to some radically new approaches. Po prevents people from common sense biases like anchoring (page 151) and conservatism (page 152), which blind us to alternatives.

Another way to provoke new ideas is to incorporate random ideas or put objects together in random combinations. You can also use po to yoke these mismatched concepts together. For example, if you're in the middle of eating a banana, you might throw together the following combination, which generates some straightforward ideas and some more creative ones:

Restaurant *po* bananas

> Let's provide mushy bananas as free baby food.
>
> Let's have a tropical theme night.
>
> Let's give our hard-working staff 10 minutes at the end of every day to "go bananas."
>
> Bananas get sweeter as they ripen. Banana lovers can choose when to eat a banana to get the sweetness they want. Let's give our customers a similar ability to choose the sweetness of their desserts using a five-point scale when they place their orders.

To get random input for a provocation you can look around you, flip open a dictionary and point to a word, eavesdrop on a conversation between strangers, or watch 10 seconds of television.

Solve the Opposite Problem

Sometimes you can gain new insight on a problem by reversing it—in other words, trying to do exactly the opposite of what you really want to accomplish.

For example, imagine you're dealing with a rash of dissatisfied customers at your customer service call center. You try to patch the problem, but a month later your clientele is still as annoyed as ever. Using reversal, you ask yourself the opposite question: "How can I reduce customer satisfaction?" Here are some ideas:

- Stop answering the phone. Don't bother returning messages.
- Put customers on hold for long periods of time. Play offensive music.
- Instruct customers to contact someone else in a different department.
- Provide meaningless advice based on poor product knowledge. If you can't answer a question, invent something that sounds technical.
- Speak quickly and mumble.
- Keep to a strict time limit. Usher the customer off the phone when their time is up, whether or not they've received the answer they want.
- Treat customers rudely and abusively. Blame them, insult them, and ridicule their questions.

Fun isn't it? As you can see, these bad ideas suggest obvious areas that you can investigate and improve to achieve the real goal. For example, are calls being returned promptly? Are customers being transferred needlessly? Are little details like hold music aggravating the problem?

Prompt Ideas with SCAMPER

When your ideas start to fizzle out, you can use the SCAMPER system developed by Bob Eberle to trigger a few new ones.

Essentially, SCAMPER is a checklist of transformations—ways you can alter an existing object or idea to come up with something new. By applying SCAMPER to your problem, you'll force yourself into creative new directions.

The following table puts SCAMPER to work. It lists the seven ingredients of SCAMPER, each of which corresponds to a letter in the word, describes the questions they prompt, and shows an example of the ideas they might turn up when applied to the floundering restaurant example you considered earlier.

Letter	Concept	Questions to Use	Restaurant Example
S	Substitute	What can I substitute to make an improvement? Can I replace people, components, materials, or processes with something different?	Replace the dinner menu with a selection of drinks and tapas.
C	Combine	Can I add something else to create something new? Can I combine objects, goals, or ideas?	Add live jazz music to transform the restaurant into an entertainment center.
A	Adapt	Can I take a solution from somewhere else, and adapt it to this one? Is there a parallel between this situation and something else?	A nearby independent bookseller narrowly avoided bankruptcy by setting up an e-commerce Web site. The restaurant could develop its own Web presence, such as links to a local foodie review site.
M	Modify	Can I change part of my object or process to make it better?	To avoid waste, remove poor sellers and dishes that need to be prepared a day in advance.
P	Put to another use	Can I take things as they are now, and use that in another way or to tackle a different problem?	Rent the dining room out as a community center, and use the kitchen as a catering business.
E	Eliminate	Will removing something streamline my problem? How will I compensate, and will that open new possibilities?	If we had no servers, we could save money and offer a buffet service.
R	Rearrange	Can I reverse or rearrange the concept I already have? Can I change the order of steps in my process?	Change the focus from a restaurant with dancing to a nightclub with food. Include the cost of food in the price of admission, and make money selling alcohol.

The Six Thinking Hats

The six thinking hats is a method created by Edward de Bono to force you to look at the same situation from several perspectives. The following list details the six hats, and provides, in parentheses, an image that can help you remember each one:

- **White Hat (a blank sheet of paper).** This hat represents objective thinking that focuses on facts and figures. Use white hat thinking when you need to ground an argument by consulting the underlying data.

- **Red Hat (a burning fire).** This hat represents emotional and intuitive thinking. Logic isn't required. Red hat thinking can praise or criticize an idea based on raw, subjective feelings.

- **Black Hat (a judge's robe).** This is the hat of cautious judgment. Use black hat thinking for a dose of critical logic, which can point out what an idea lacks or where it doesn't fit the facts.

- **Yellow Hat (the happy sun).** This is the hat of praise. It focuses on the reasons why an idea will work and how it can provide benefits. Yellow hat thinking can sometimes pull the good news out of a seemingly dire situation.

- **Green Hat (a growing plant).** This hat represents creativity. It provides provocations, new ideas, and outrageous alternatives, with no effort to criticize or evaluate the merits of these ideas. Use green hat thinking to shake things up and set off on a new direction.

- **Blue Hat thinking (the sky).** This hat represents the big picture. It focuses not on the problem at hand, but the *way* people are approaching that problem. For example, you can use blue hat thinking to set goals, determine how a meeting should be held, and suggest which hat should be worn at a particular time to advance the discussion.

Six-hat thinking can be used on an individual basis or in a group. When you're wearing the hats on your own, you'll need to try on each hat one after the other, which can be a bit time consuming.

The six thinking hats really shine in group settings. Usually, everyone dons the same hat at the same time, and the discussion in a typical meeting travels through a sequence of different hats. For example, you might don your blue hats to describe the problem and formulate a strategy of how to approach it, switch to the red hats to collect visceral opinions about the issue, grab the green hats to generate new ideas, and then continue to the white, yellow, and black hats to weigh and adapt these ideas into a final solution. (Alternatively, you can assign different hats to different people, which sounds like fun but generally makes it more difficult to build on each other's ideas.)

The six thinking hats work well in group settings because they give people the freedom to express different views without having to defend them personally. For example, the black hat allows criticism of an embarrassingly bad idea, the red hat gives permission to vent about recent frustrations, and so on. The hats also encourage people to try out roles they don't usually embrace, such as praising an idea that would normally be viewed with suspicion (yellow hat) and suggesting an idea that would ordinarily meet with ridicule (green hat).

> **Tip** If you're interested in honing your creative thinking, you don't need to stop with the suggestions shown here. Check out one of the popular books on the subject, such as Michael Michalko's *Thinkertoys* (Ten Speed Press, 2006) or one of the many books by self-proclaimed creativity expert Edward de Bono, who first coined the term *lateral thinking* and who devised the six thinking hats described here.

8 Your Personality

So far, you've spent seven chapters digging into the tangled mass of neurons that comprises *you*.

For most of this time, you've focused on the common characteristics that every normal brain shares. You've learned that brains crave food and sleep. You've seen how they perceive things and remember them. You've also seen how they use emotions to drive you and leaky logic to explain the world. But while these topics are undeniably fascinating, they do little to separate *your* brain from that of a teenage parking attendant, a professional wrestler, or a theoretical physicist. To explain the constellation of attitudes, traits, and temperaments that distinguish individuals, you need to consider something that's much more difficult to pin down: *personality*.

From a neurological point of view, personality is a phenomenon that's created by the interaction of a huge number of different brain parts. On some level, personality involves all three pounds of soggy brain jelly. In fact, it's more than likely that personality is nothing but a catch-all label to describe the idiosyncratic way that each brain juggles its perceptions, memory, emotions, and reasoning when making life decisions.

But here's the important part—although personality is a fuzzy concept, it's not a dead end in your brain exploration. There's good reason to believe personality is biologically rooted and difficult to change. And while no two people share exactly the same personality, the same personality "themes" crop-up across the world and throughout history. Put these two facts together, and you'll see why exploring the quirks of your own personality is worth the time. First, it gives you a tool to make sense of your past. And second, it helps you chart out your future, so you can pursue the people, places, and activities that fit your unique character.

The Building Blocks of Personality

For nearly 100 years, psychologists battled over the key components of personality and different theories proposed different ways of splitting people into categories. Famous classifications included *extraverts* versus *introverts*, hard-driving *Type A* personalities versus more leisurely *Type B* personalities, and intuitive *perceivers* versus more analytical *judgers*. These theories rarely agreed with each other, failed to capture the full spectrum of personality, and inspired a paper mountain of personality tests.

The solution to this morass of conflicting analyses started with something called the *lexical hypothesis*, which a few deep thinkers suggested in the early 1930s. The lexical hypothesis suggested that researchers could find the fundamental ingredients of personality by analyzing the thousands of personality-describing adjectives in the English language. After all, language provides the framework we use to understand the world around us. It's also a distillation of the observations made by countless generations of real people.

But there was a problem. After consulting quite a few dictionaries, researchers started out with a dizzying 18,000 adjectives. Analyzing them would require some serious work. So instead, psychologists took a collective break, grabbed some coffee, and returned to the easier studies of the day, which usually involved ringing tiny bells before giving a pigeon its dinner.

The problem was revisited several times and finally attacked with a device that had the potential to work much harder than the average research psychologist: the computer. Eventually, the original list of adjectives was boiled down to a combination of just five factors, called the *five factor model* (or sometimes the "big five," as though it were a meeting of mafia bosses).

To perform this task, researchers analyzed thousands of surveys using some surprisingly hard-core statistics. These surveys asked recipients to classify their personality or other people's personalities by ticking off adjectives on a list. If certain words tended to be ticked off in combination, researchers concluded that they were part of a related personality factor. For example, statistical analysis found that the words *outgoing*, *social*, and *gregarious* were usually chosen together. Although these words don't mean the same thing, this correlation suggests these concepts can be combined into a broader personality factor—in this case, *extraversion*. Similarly, people who choose these words usually avoid others, such as *shy*, *quiet*, and *reserved*. Thus, these words are the opposite side of the same concept. They indicate an extreme lack of extraversion.

Incidentally, the five factors that researchers settled on were *openness*, *conscientiousness*, *extraversion*, *accommodation*, and *neuroticism*. To describe an individual's personality using the five factor model, you score each of these factors separately.

How Realistic Is the Five Factor Model?

Now that you know how the five factor model was created, you can pick out some potential holes in its logic:

- **It's all based on words.** Perhaps there are deeper personality elements that we haven't noticed or haven't named. If so, any analysis that starts with the English language is doomed to leave something out.

- **It's based on Western societies.** Different cultures have different values, which shape how people perceive certain characteristics. For example, a person who's wildly extraverted by one culture's standards might seem distinctly average by another's.

- **It involves questionnaires.** Psychology is the only field of science where significant studies rely on the honesty, accuracy, and attention of ordinary people. Some problems are inevitable—some people might be too bored to finish the test, too embarrassed to admit they don't know what the word *gregarious* means, or too drunk to find the correct end of their pencil. Psychologists hope that over time, with different questionnaires and large numbers of people, these many discrepancies average themselves out. However, it's possible that the five factor model reflects how survey participants think about personality as much as it reflects their actual personality traits.

- **The associations are based on most people, not all people.** In other words, you might be an unusual person who's highly private but still craves social contact. Because the five factor model reflects the more common combination—social gadflies who share their lives around like an open e-book—it can't capture the full essence of *you*. In fact, the extraversion factor wraps a whole series of interrelated traits that are closely correlated in many people, including assertion, talkativeness, and confidence.

- **The focus on five traits masks other issues.** It's easy to forget how people adapt their personality in different contexts. For example, Bill may be the jiving life of the party at his cousin's wedding, but he downshifts for his day job at the mortuary.

At first glance, all these points would seem to strike serious blows against the five factor model. However, the picture isn't as grim as it seems.

First, you need to realize that the model isn't trying to pass itself off as a brain *fact*. It doesn't suggest that there are five distinct regions in the brain that generate different types of personality. It doesn't intend to supply a five-ingredient recipe that can completely capture your personality. Rather, the five factor model is a psychological tool. It gives you a powerful way to look at, size up, and talk about different personalities, without slotting them into limiting categories.

 Note As with many models, there's a tension between capturing all the details and creating useful generalizations. In other words, reducing personality to five dimensions is obviously a simplification, but it's a *useful* simplification. In the real world, the five factor model can often pull out traits that predict life outcomes like juvenile delinquency or academic success. It can also make generalizations about the personalities of people in a particular career. (For example, one study shows that flight attendants tend to have high scores in extraversion, openness, and neuroticism.)

The five factor model gets a lot of love because it resolves the tensions of the dozens of personality models that existed before. The five factor model also has a few impressive wins on its side—for example, the same five factors have turned up in many different analyses and in many cross-cultural studies (although the individual factors are often shaded with slightly different meanings or given different emphasis).

There's also some evidence that certain traits correlate to identifiable neurological factors—for example, higher levels of certain neurotransmitters, increased activity in certain areas of the brain, and so on. However, it's hard to pin down whether these correlations are causes or just side effects, and so we won't spend much time on them in this chapter. Ultimately, it's up to you to decide whether personality research is a legitimate science or a creative endeavor with a generous heaping of warmed over statistics.

A Personality Test

Before you take a closer look at the five factor model, it's time to see where your own personality falls on the scale.

 To download a printable copy of this test, visit the "Missing CD" page for this book at *www.missingmanuals.com*. If you prefer to take an online test that does the scoring for you automatically, you can find a similar tool online at *www.outofservice.com/bigfive*.

The following quiz lists 50 statements. Next to each one, write a number from 1 to 5, as follows:

> 1: Very Inaccurate
>
> 2: Moderately Inaccurate
>
> 3: Neither Inaccurate nor Accurate
>
> 4: Moderately Accurate
>
> 5: Very Accurate

The goal is to describe yourself as you are now, in relation to other people you know of roughly the same age—not how you wish to be in the future.

1. _____ I don't mind being the center of attention.
2. _____ I feel little concern for others.
3. _____ I'm always prepared.
4. _____ I get stressed out easily.
5. _____ I have a rich vocabulary.
6. _____ I don't talk a lot.
7. _____ I make people feel at ease.
8. _____ I leave my belongings lying around.
9. _____ I'm relaxed most of the time.
10. _____ I have difficulty understanding abstract ideas.
11. _____ I feel comfortable around people.
12. _____ I insult people.
13. _____ I pay attention to details.
14. _____ I worry about things.
15. _____ I have a vivid imagination.
16. _____ I keep in the background.
17. _____ I sympathize with others' feelings.
18. _____ I tend to make a mess of things.
19. _____ I seldom feel blue.
20. _____ I'm not interested in unrealistic ideas.
21. _____ I start conversations.
22. _____ I'm not interested in other people's problems.
23. _____ I follow a schedule.
24. _____ I'm easily disturbed.
25. _____ I have excellent ideas.
26. _____ I keep quiet around strangers.
27. _____ I have a soft heart.
28. _____ I often forget to put things back in their proper place.
29. _____ I seldom get mad.
30. _____ I don't have a good imagination.
31. _____ I talk to a lot of different people at parties.
32. _____ I'm not really interested in others.
33. _____ I like order.

34. ____ I get irritated easily.

35. ____ I'm quick to understand things.

36. ____ I don't like to draw attention to myself.

37. ____ I take time out for others.

38. ____ I shirk my duties.

39. ____ I rarely have mood swings.

40. ____ I try to avoid complex people.

41. ____ I am skilled at handling social situations.

42. ____ I am hard to get to know.

43. ____ I do things according to a plan.

44. ____ I grumble about things.

45. ____ I love to think up new ways of doing things.

46. ____ I find it difficult to approach others.

47. ____ I show my gratitude.

48. ____ My office or workspace is cluttered.

49. ____ I feel at peace with the world.

50. ____ I avoid difficult reading material.

Scoring the Test

Once you've finished, it's time to find out how your brain did. To score the test, add up five numbers, each of which corresponds to one of the five factors. Each factor score ranges from a maximum of 20 to a minimum of -20. If you score a 0, you fall dead in the middle.

To calculate your *extraversion* score, add 1, 11, 21, 31, 41 and subtract 6, 16, 26, 36, 46.

To calculate your *accommodation* score, add 7, 17, 27, 37, 47 and subtract 2, 12, 22, 32, 42.

To calculate your *conscientiousness* score, add 3, 13, 23, 33, 43 and 8, 18, 28, 38, 48.

To calculate your *neuroticism* score, add 4, 14, 24, 34, 44 and subtract 9, 19, 29, 39, 49.

To calculate your *openness* score, add 5, 15, 25, 35, 45 and subtract 10, 20, 30, 40, 50.

To find out what these numbers actually mean, keep reading.

Dissecting Your Personality

In the following sections, you'll analyze each score separately. To get an overview of where you fall, you can fill your results into the scoring sheet starting on the next page (also available for download from *www.missing-manuals.com*).

On either side of the scale for each personality factor are some commonly used adjectives for people with high or low scores. For example, high extraversion scorers are more likely to be described as talkative and assertive.

> **Note** Each of the five factors represents a continuum. The "good side" and "flip side" lists you see here define the extremes. Most people fall somewhere in the middle and have characteristics from both sides. However, they usually have more of one side than the other. It's also important to remember that your position can shift as you age, and may change based on your current mood.

Extraversion

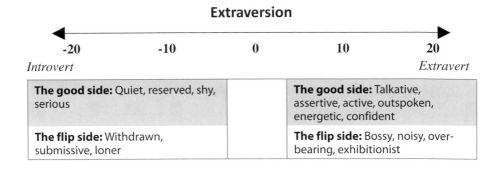

-20	-10	0	10	20
Introvert				*Extravert*

The good side: Quiet, reserved, shy, serious		**The good side:** Talkative, assertive, active, outspoken, energetic, confident
The flip side: Withdrawn, submissive, loner		**The flip side:** Bossy, noisy, overbearing, exhibitionist

Accommodation

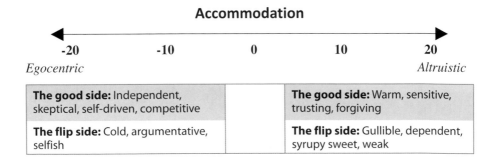

-20	-10	0	10	20
Egocentric				*Altruistic*

The good side: Independent, skeptical, self-driven, competitive		**The good side:** Warm, sensitive, trusting, forgiving
The flip side: Cold, argumentative, selfish		**The flip side:** Gullible, dependent, syrupy sweet, weak

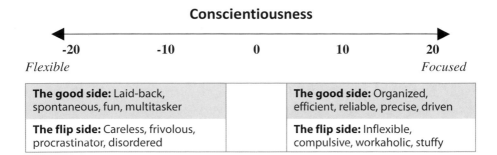

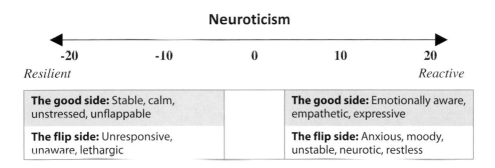

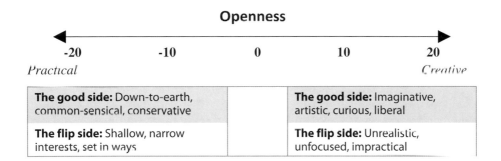

The following sections get into much more detail about what your five personality scores mean. They also provide some advice for living life with the personality you've got.

Extraversion

The extraversion factor indicates how strongly you crave interaction with the outside world. Extraverts desperately need to be wired into other people and social situations. They feed on parties, meetings, and chit-chat. If you wear cowboy boots to the dentist's office and you don't live in Texas, you're a clear-cut extravert. By comparison, introverts are quieter, low-key people who seek out individual activities. They're often highly independent and are more comfortable alone. If you envy the relaxed solitude of your backyard garden gnome, you're probably an introvert.

It's obvious that all humans need to strike a balance between being on their own and hanging out with other people. The extraversion factor of personality reflects each person's personal compromise between these two needs. At some point, even the most outgoing person needs to withdraw for some one-on-zero time to rest and recharge. Introverts seem to have a much lower threshold—they're ready to pull out of the social world much sooner, and take longer to get ready for the next encounter.

 Note It's theorized that introverts have a higher baseline state of arousal. This means, it takes less social interaction to stimulate them. On the other hand, extraverts need far more stimulation to reach the same level of arousal and get their social needs met.

If you scored high…	For you, social interaction is a source of positive emotions and energy. You'll be happiest if you can choose social roles where you're in the limelight. The more extraverted you are, the more likely you are to crave attention and play a leadership role. Sales people, actors, politicians, and managers are archetypal extraverts.
If you scored low…	For you, social interaction can be exhausting—it's usually an energy drain. You'll be happiest working independently, with minimum noise and distraction. The reclusive writer is an obvious introvert.

To compensate for a low score…	Introverts will never become extraverts, but introverts seeking change can get used to levels of social stimulation that would normally send them dashing to the peace and quiet of a dark cave. Simply being in a social situation isn't enough—introverts need to practice engaging with it. Books on assertiveness, small talk, and public speaking may help. And don't be embarrassed to rehearse what you want to say in advance.
To compensate for a high score…	Sometimes, extraverted people can overpower others and miss out on the subtler rewards of quiet time. To cultivate your introverted side, schedule an hour a day for daily reflection (for example, start writing a journal). In interactions with other people, let others talk first. When dealing with more introverted people, home in on what they say and ask follow-up questions to pull them into the conversation. And in large group settings, let everyone have a turn talking before you jump back into the spotlight.

Accommodation

The accommodation factor (also called *agreeableness*) indicates how likely you are to cooperate with others, even at the expense of your own needs and desires. Highly accommodating people value social harmony and are willing to compromise to get it. Low accommodating people are focused on their own personal viewpoint. They value logic, and trust that it will show the inarguable correctness of their every opinion. They aren't necessarily selfish, but they aren't willing to relinquish their view of the world to get along. They're also less likely to extend themselves to help others, and have a healthy level of skepticism about other people's motives. If you say "but" more than "yes," and have chronic headaches from frequent eye rolling, this describes you.

If you scored high…	You're an adapter. This gives you a remarkable ability to smooth over rough edges, negotiate, and make peace. You'll be effective in a social role, but you may need to catch yourself before you bend too much, trample on your sense of self, or end up overly dependent on a romantic partner. Adapters often end up in nurturing or caregiving roles.
If you scored low…	You're a challenger. When dealing with others, expect a little friction. Low accommodation is good for situations that need tough thinking or objective logic. Challengers often end up in roles where they can impose their opinions on others—for example, as restaurant critics.
To compensate for a low score…	Become aware of the challenging signals you send, such as using abrasive words, making summary judgments, dismissing other opinions out of hand, and interrupting (the most counter-productive challenger habit). If you curtail these practices, you'll make room for other people to express themselves and join in the decision-making process. That way, they'll be more likely to support the final decision.
To compensate for a high score…	Focus on yourself. When dealing with other people's needs, remind yourself of your own opinions and desires. This internal monologue may keep you from stretching too far.

Conscientiousness

The conscientiousness factor indicates how easily you can control your own impulses and implement a plan. Highly conscientious people are focused, methodical, and prepared. They make plans in advance, implement them with careful attention to detail, and generally end up where they want to be. Conscientious people are on time, on budget, and always match their socks. By comparison, people who have low conscientiousness scores are harder to motivate and more easily distracted. They aren't particularly enamored by goals, but they can disengage from complex tasks and relax at the end of the day. They also have the benefit of being flexible in the face of changing circumstances. Their underwear may be inside out, their keys may be locked in their car, and their pet python might be slipping out an inadvertently open window, but they feel fine.

Again, there's tantalizing evidence that conscientiousness is rooted in specific features of the brain. It's possible that the reward circuitry you explored in Chapter 6 is balanced slightly differently in more conscientious people, making immediate rewards less seductive and long-term goals more compelling. Or, highly conscientious people just might be better at harnessing their prefrontal cortex (page 148) to predict future outcomes.

If you scored high…	You need order, and with it you can excel. However, don't try to live life like a person with a low conscientiousness score. If you don't plan ahead and neaten up your house, you're likely to feel deeply disturbed. Seek structured environments that you can control so you'll always be on top of your game.
If you scored low…	You're a born relaxer. You might have an easier time enjoying life, reducing, shifting gears, and dealing with change, but you'll need all the help you can get for long-term planning.
To compensate for a low score…	Use daily to-do lists and reminders (page 118) to keep yourself on track, and set rewards for reaching your goals. Use priorities to separate the critically important from the merely interesting. Streamline your work environment so it doesn't contain distractions, and make rules (for example, no wandering out of the office for an impromptu game of ping-pong).
To compensate for a high score…	The dark side of high conscientiousness is workaholism and stress. Cultivate an end of day routine that gets you away from your work and prepares you to relax. (One way to do so is to end each day by adding all your unfinished tasks to tomorrow's to-do list.) If possible, delegate work to others, both in your career and at home. (For example, if you need a supremely clean abode, hire a cleaning service.) Lastly, schedule in a slot for activities that aren't goal-directed, such as a leisurely conversation with friends.

Neuroticism

The neuroticism factor (also known as the *need for emotional stability*) indicates a little more than you'd care to admit about your ability to deal with negative emotions. People who score high on the neuroticism scale are emotionally reactive. They respond with negative feelings to events that others might ignore. They're quicker to feel anxiety, anger, and depression, and these feelings often linger for long periods of time.

To a high neuroticism scorer, ordinary situations often seem threatening, and minor frustrations can quickly become maddening. (Worrisomely, this is the portion of the category that customer service people tend to score in.) By comparison, people who score low on the neuroticism scale are even-tempered and relaxed. That doesn't mean they're any happier, but they aren't likely to pin you under a chair when a jar of cheese spread won't relinquish its lid.

 Note Neuroticism describes your susceptibility to *negative* emotions. It doesn't reflect how you respond to positive emotions. Incidentally, frequent feelings of positive emotion are most closely correlated with high extraversion scorers. (Of course, this relationship is influenced by the culture in which you live. In a society that prizes discretion and frowns on extraversion, extraverts will feel sorely out of place.)

If you scored high…	To be at your most decisive, accurate, clear-thinking best, you need to be free from the cloud of anxiety. Avoid situations, and people that trigger bad feelings, because these feelings can easily overshadow the entire day. You'll perform best in neutral work environments that don't have strong emotional cues. Environments that feature shouting people, flashing lights, and a din of noise—think of the exchange floor where stock traders work—are less suitable.
If you scored low…	Negative emotion rarely interferes with your day-to-day life. However, recognize that your perception is shaped by the things that *you* respond to, and as a result you're likely to miss the details that annoy others.

To compensate for a low score…	Although you're able to navigate trying times, you also have an emotional blind spot. You'll need to work at anticipating potential sources of strife. (Hint: Bad ideas include whistling the entire *William Tell Overture* on the subway, leaving last week's yogurt to grow exotic molds in the company refrigerator, or pausing thoughtfully when asked "Do these pants make me look fat?")
To compensate for a high score…	Short of medication, there's no quickfire way to become less responsive to negative triggers. Instead, work to reduce stress (using the tips on page 138) and practice reframing grating situations with a positive explanatory style (see page 144).

Openness

Openness is the vaguest of the five factors, and the one that changes the most in cross-cultural studies. It's sometimes described as originality, openness to experience, or intellect. It attempts to capture your intellectual curiosity about life and the universe. People who score high on the openness index are described as open and creative. They seek out novelty, hold unconventional beliefs, and spend more time in introspection. Most appreciate art, literature, and culture.

By comparison, people who score low on the openness scale are more down-to-earth. They have straightforward, conventional priorities, prefer the simple to the subtle, and are more comfortable with familiar places, people, foods, and ideas. They're also squarely focused on the practical side of things, which often leads them to dismiss subjects with no obvious applications (for example, abstract art, medieval philosophy, and theoretical physics). If you often imagine what it would be like to ride a llama through Grand Central Station, you're probably a high-scoring creative type. If you're irritated by even the possibility that someone would mentally cook up that idea, and mostly concerned about how a runaway llama might affect your morning commute, you probably rank lower on the openness scale.

If you scored high…	As a creative person, you need to feed your hunger for novelty—new ideas and experiences. Without them, your brain will wither from boredom. Your strength is your ability to think creatively and innovate, but it takes practice to whittle new concepts into something usable.
If you scored low…	As a practical person, you need to see a concrete rationale to a project before you can commit your whole brain to it. Your strength is your ability to take an outlined idea and implement it successfully. You'll work best by following established practices and patterns. Your preferred strategy is to incrementally refine an idea, rather than replace it with something new, risky, and untested.
To compensate for a low score…	A little creativity can go a long way to starting a process or getting you out of a rut. Use the creative thinking techniques in Chapter 7 (page 171) to help you out.
To compensate for a high score…	From time to time, you'll need to rein in your flighty side. To do so, practice planning and create a to-do list with concrete deadlines and rewards. For true problem solving power, consider teaming up with a low-openness scorer.

 Note Don't assume that open-minded free thinkers are smarter than more practical people with lower openness scores. In traditional measures of intelligence, both types of people do equally well. It's no stretch to see that society needs a combination of innovative high-openness scorers and more prudent low-openness scorers to run smoothly.

The Personality Fit

You can think of the five factors as one way to take a snapshot of your personality. They won't capture your many quirks (or your dashing good looks), but they will show the outlines of five key personality characteristics.

When you start affixing labels and numbers to something as nebulous as the brain, it's easy to get carried away. Nowhere is this more obvious than in the field of career planning, which attempts to slot people into specific jobs based on the strengths and quirks of their personalities. As you'll see, career testing is often sold for more than it's worth.

Career Testing

Long loved by high school counselors, the typical career test clarifies absolutely nothing. Career tests are famous for brilliant pronouncements like this: *Your nurturing side suggests you'd make a fine doctor, veterinarian, or housewife. Your need for order suggests you'd enjoy life as a judge, accountant, or sanitation worker.*

 Tip Career tests aren't all bad. They're great as a brainstorming tool to help you identify positions that could be a good fit, and get you thinking about why certain roles appeal to you. Just don't expect them to unearth your secret dream job.

The problem with mapping personalities to careers is that most careers have room for a range of different roles. Although some professions require extreme personality types (for example, the prospects are dim for introverted break dancers and low-conscientiousness house cleaners), most are surprisingly flexible. For example, it's obvious that the self-directed focus that's needed for a film critic, computer programmer, or research scientist makes these good positions for introverts. However, extroverts will be just as happy if they have the freedom to express their social side—say, interviewing actors and discussing movies at a film festival (movie critic), leading a team of code warriors into battle (programmer), or collaborating with colleagues and presenting new techniques and ideas (scientist). Personality analysis reveals that all these professions are more commonly occupied by introverts, but that fails to account for the perfectly happy extroverts who have carved out a slightly different niche in the same world. Similarly, job satisfaction also relies on many other factors, such as the type of work you're doing, its rewards (in money and prestige), your time commitment, the attitude of the people you work with, and how high your boss scores on the neuroticism scale.

Rather than attempt to use personality analysis to steer you toward the right jobs, hobbies, and interests, you'll have an easier time using it to illuminate why you like certain things, why you resent others, and what's still missing. For example, if you score high on the openness scale, you might grow restless in an otherwise excellent job because you aren't getting the novelty you crave. Possible solutions include changing your role and responsibilities at work, switching careers, or picking up a new hobby in your free time. The decision is yours to make, but it's the five factor model that helps you pin down the problem.

 Note As a general rule, if your life isn't in harmony with your personality, it's your life that needs to change. Although you can learn to compensate for extreme scores, your inner core is like a monolithic glacier. It may drift with the years, but you won't be able to push it very far.

Flow

Now's a good time to review your freshly calculated personality scores and think about how they line up with your day-to-day life. As you learned in Chapter 6, you can't find happiness by pursuing pleasure. However, you just might enter a state of deeper satisfaction when your life is in harmony with your personality.

This state, sometimes called *flow*, is achieved when you're challenged but not worried, tested but not frustrated, engaged but not fatigued, and confident but not bored. When you're in a state of flow, you're forced to grow even though you don't feel out of place. Most importantly, you have the chance to use all aspects of your personality. If everything goes right, this leaves you (cue the new age music) fully actualized and at one with the universe.

The flow metaphor underlies everything from a star athlete being "in the zone," to an artist having a burst of creativity, to a video game wizard driving Pac Man through a 24-hour quest for colored dots.

 Tip For more on flow and some inspiring ideas on how to get into this wildly productive state, check out any of the books with that word in their title by Mihály Csíkszentmihályi (just don't try to pronounce his last name).

Picking a Place to Live

One of the many intriguing relationships in personality research is between people's personalities and the places they live. For example, the personality testing at *www. outofservice.com/bigfive* has found that people in places with high precipitation have higher scores on the neuroticism index, places with high population densities have higher scores in openness and accommodation, places with greater ethnic diversity have higher scores in openness, and so on.

There's no easy way to say whether this relationship shows how the environment shapes personality, or if people choose environments that are suited to their personalities (or both). For example, it's plausible that people with high openness scores favor diverse, bustling, novelty-rich metropolises. Alternatively, living in these places may cause people to become more open. Or, there may be a more subtle effect underway based on the history of a particular location and the attitudes around it. For example, the urban pride of city-dwellers may lead them to see themselves as more open-minded than they really are.

In any case, one thing is clear. The fit between personality and environment is just as important as the fit between personality and career types. If you're living in a place you don't like—for example, feeling lost and alone in the wide open countryside, suffocating in the rushing crowds of a city, or feeling emotionally empty in the dreary sameness of suburbia—don't stay there. Find the right home and pack your bags.

9 The Battle of the Sexes

To state the obvious: men and women have different anatomical endowments. Some people downplay these differences, others emphasize them, and a lot of us obsess over them, despite the fact that we have absolutely no background in biology. But for neuroscientists, who are more interested in studying the sex-specific charms of the brain than those of the body, the allure wears off fast. Trying to use the brain to unravel the mystery of the sexes is a sure way to lose grants, alienate potential dates, and wear out your MRI machine.

It's not that male and female brains don't have differences—they clearly do. However, determining the **significance** of these differences is another matter entirely. In fact, it's a challenge that's puzzled researchers, taken down a Harvard president, and landed countless neuroscientists on the living-room couch.

In this chapter, you'll see what all the fuss is about as you hunt for sexual differences in the brain. What you'll learn is compelling, controversial, and often inconclusive—but it just might give you a new perspective on the person you've sworn to spend your life with. Just as fascinating as the question of what makes us so different is the puzzle of what keeps us together (at least long enough to make promises, babies, and mortgage payments). In the second half of this chapter, you'll try to answer this question by considering what happens to the brain when it falls in love.

Gender in the Brain

In order to understand how your brain is shaped by your inherent male- or female-ness, you need to know a bit more about how babies are made. (No, not *that*.) What you need is a refresher in the high school science that explains how a sperm and an egg partner up to build boys and girls.

As you'll see, the same biological processes that shape the sex of your body also leave a subtler imprint on your brain. Few agree on the exact effects of these processes. But if we're ever going to stop throwing chairs at each other and sit down to a sensible debate about gender, we need to start with the basics of biology.

Sex and DNA

Scientists know a fair bit about what makes new people into little boys and girls. It all starts with the compact spools of DNA called *chromosomes*. As a human, you have 23 pairs of chromosomes curled up in virtually every cell of your body. (That's 46 chromosomes in each cell, for those of you who nodded off in math class.) If you could uncurl your chromosomes and lay them out on a tabletop, they'd look like the figure shown here. (Assuming you're a guy. If you're a woman, you have two copies of the X chromosome at the bottom-right of the figure, and no Y chromosome.)

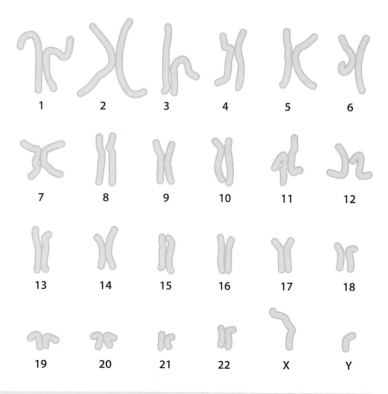

When two people get together in the interest of DNA sharing, they pass along *one* chromosome out of each pair. If you're a woman, these chromosomes are packaged up in an egg. If you're a man, they're carried along by your trusty swimmer, the sperm. When a sperm and an egg cell meet, this combination of chromosomes is reassembled into the 23 pairs needed to produce an entire human being, complete with body, brain, and adult-onset neuroses.

The Y Chromosome

The last pair of chromosomes determine sex, and scientists named them X and Y. All chromosomes contain sequences of different genes, and the Y chromosome contains a gene called SRY (which rather unimaginatively stands for *sex-determining region of the Y chromosome*).

So if you get the Y chromosome, you acquire the SRY gene, which is genetically programmed to go to work creating testicles. As a result, you'll wind up a man. Your other chromosomes have nothing to say about it.

 Note If you do get the Y chromosome, it's passed to you by your father. (Your mother has two versions of the X chromosome, so you're bound to get an X chromosome from her.) The common shorthand is to say that men are XY and women are XX.

There's an ongoing debate as to how many useful genes the stumpy Y chromosome actually has. In the past, scientists believed the Y chromosome was a genetic wasteland. However, recent studies suggest that it includes about 80 genes for male fertility, sperm production, and perhaps other biological functions. (You can compare this with the more than 800 genes on the X chromosome, which both sexes get.) This picture of the two shows why many men have chromosome envy.

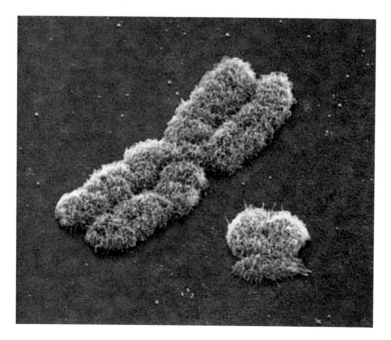

 Note The reasons why the Y chromosome is so small and stunted are a bit technical. Essentially it's gradually deteriorated over the millennia because it can't use the error-checking abilities of its more feminine partner, the X, which is probably the best metaphor for male and female behavior that you're likely to find in genetic science.

The overall significance of the genes on the Y chromosome is a mystery. What's clear is that every other chromosome is sex-independent. In other words, aside from the Y chromosome, you have an equal chance of picking up genes from your father or your mother, regardless of whether you yourself are a man or a woman. And yet some scientists point out that, statistically speaking, the Y chromosome creates a genetic difference between men and women that's roughly the same as that between a male human and a male chimpanzee—a fact which comes as distinctly no surprise to many.

Sex Hormones

The SRY gene, if you have it, will set you on an irreversible course to becoming a man. However, in order to effect this transformation, it needs the help of another player—*testosterone*.

When a male embryo reaches its sixth week of existence, the infamous SRY gene creates a protein that triggers a complex series of actions, and ultimately leads to the creation of testicles. If the SRY gene isn't at work, the embryo develops ovaries instead.

 Note You may have heard the myth that every embryo begins as a female. In reality, we begin with all the plumbing we need to go either way. Technically, a six-week old embryo is *sexually undifferentiated*. It contains basic anatomy that can be converted to male testicles or female ovaries, depending on the influence of the all-important SRY gene.

Testosterone is the key hormone for sexual differentiation. As soon as the testicles are formed, they begin to secrete testosterone, which has a range of effects across the body. Interestingly, no such hormonal helper is needed for female embryos. In the absence of testosterone, the body automatically begins building female body parts.

Besides the obvious (building our naughty bits), testosterone also passes into the brain. Humorously, it's then converted into a form of estrogen known as *estradiol*. (In other words, the hormone that shapes a male brain is the one most associated with females.) Although we can't watch estradiol work directly, scientists suspect it's up to something for the following reasons:

- The brain has specific receptors in specific regions that respond to estradiol and other hormones. So it isn't ignored.

- In the lab, dosing certain brain regions with estradiol triggers visible changes. For example, scientists who extract tissue from rat brains find that estradiol works like Miracle-Gro, causing the neurons to sprout bushy dendrites that are perfect for making new connections.

- In studies of other animals, researchers consistently find that testosterone wires the brain for sex-specific behaviors like birdsong, building nests, and mounting potential mates. If you change the level of testosterone shortly before or shortly after birth, you end up with males that show more female behavior, or vice versa. These effects have been studied in a wide range of animals, including rats, hamsters, ferrets, finches, pigs, dogs, and sheep.

 Note Before you make too much of the huge library of animal studies, it's important to remember that in most animals sex-specific behavior is highly stereotyped. In other words, their brains include hard-coded patterns of behavior that simply need to be activated at the right time. Primates (and humans in particular) behave quite a bit differently. Our brain includes a huge cerebral cortex that prefers to learn new behaviors rather than follow preset instincts.

Fun Facts

Can Raging Hormones Account for Sex Differences?

With the soaring levels of male testosterone, you might assume there's no need to hunt for sex differences in the brain. After all, isn't testosterone more than enough to account for belligerent male behavior?

In fact, the idea that testosterone fosters aggressiveness is a widely held but poorly supported myth. It's clear that testosterone is required to support certain behaviors, which is why castrated bulls don't get quite as worked up as they used to. It's also true that violent criminals have higher testosterone levels (in prison, anyway). However, it seems more likely that testosterone plays a *facilitative* role rather than a causal one.

For example, the body might ramp up testosterone production to prepare to meet a challenge or fend off competition. Men who often battle it out (either by choice or by necessity) will then have higher levels of testosterone. However, studies that try to create raging man-monsters by injecting volunteers with extra testosterone find that it actually has little effect.

Testosterone doesn't exert its brain-shaping power continuously. Instead, the male body uses testosterone to set development on its way at three crucial periods: in the womb before birth, in the first few months after birth, and in puberty. At these times, testosterone is said to have an *organizational* role, because it physically alters the body (or "organizes" it).

After this point, testosterone plays a host of lesser roles—it supports a healthy sex drive, helps maintain muscle mass and bone density, and so on. It no longer sculpts the body and brain.

Testosterone isn't the only hormone at work in the body, but it's the key player in the fetal world. The fetus uses a special protein to disable estrogen— without it, both sexes would be exposed to the large quantities passed through the mother's bloodstream. After birth, this protein is no longer needed. That means the female body can begin using estrogen to trigger sexual development, much as the male body uses testosterone.

Male Brains and Female Brains

So far, you've seen how testosterone puts its greasy thumbprint on the brain. You've learned that testosterone and other hormones are able to change the size, density, and connectivity of different brain regions, but you haven't considered where these effects come into play.

Studies don't always agree on these points. However, here are some commonly cited findings:

- Male brains are marginally bigger, even when body weight is taken into account. However, females have more densely packed neurons in certain brain regions.

- Female brains reach their maximum size earlier. Different parts of the brain also appear to develop at subtly different rates.

- A specific region of the *hypothalamus* is larger in men, although no one knows exactly what this region does.

 Note It's no surprise that the hypothalamus is different in male and female brains. As you learned in Chapter 1, the hypothalamus controls the pituitary gland, which is the brain's built-in hormone dispenser. The hypothalamus releases growth hormones on a specific schedule, which depends on a person's sex. It also releases other hormones that instruct the testicles or ovaries to produce more sex hormones (like testosterone and estrogen) to push the body to develop.

- The *suprachiasmatic nucleus* in the hypothalamus is a different shape in male brains (where it's spherical) than in female brains (where it's more elongated). The suprachiasmatic nucleus is the body's timekeeper, which you met in Chapter 3 (page 46). Presumably, this different shape has something to do with the different rhythms in male and female bodies. After all, the hypothalamus is in charge of triggering female ovulation.

- Some studies have found that female brains have a thicker *corpus callosum*, which is the cable that connects the two halves of the brain. Although this finding is contentious, it hasn't stopped some scientists from suggesting that this possible difference gives women a better ability to integrate different skills.

Most of these points are the differences of averages, which means that there's considerable overlap between male and female brains. Few of the variations are as clear cut as the differences we can find in other animals. Furthermore, some of these differences could be the difference of *behaving* male or female rather than *being* male or female. In other words, it's possible to make an argument that living life as the aggressive, dominant male society expects you to be causes your brain to change, much as a life of constant navigation beefs up the hippocampus of London taxi drivers. However, the most likely culprit for these differences is testosterone and other sex hormones.

Are Gender Differences Real?

It seems like an absurd question. After all, a quick look at the biological plumbing turns up a few unambiguous differences. But when dealing with people's *behavior*, the answer isn't nearly as straightforward.

Initially, the case for sex-specific brain differences seems to be on solid ground. Aside from the less than one percent of individuals who are born with ambiguous sex organs, people can be divided into two clear groupings—male and female—and these groupings hold in all the cultures of the world and throughout history. This is quite different than the situation with race. As you learned on page 153, the way we define races overlaps sloppily (at best) with the real genetic differences between groups of people. Tradition, migration, and the politics of power influence how races are defined—and the races themselves change, shaped by generations of romantic hookups.

However, the study of the sexes quickly runs aground when it tries to make solid links between differences in behavior and fundamental human nature—in other words, when it attempts to argue that men are genetically conditioned to act like men and women are genetically conditioned to act like women. Consider this list of average sex differences, all of which are well established in many studies:

- Men are more likely to perform outward aggressive acts (like throwing heavy objects).

- Women are more likely to show empathy (whether it's by sharing toys as a child or recognizing the emotional expression of a face).
- Men perform better at certain tasks that require spatial skills (such as mentally rotating a shape and making sure a projectile hits a target).
- Women perform better at certain tasks that require verbal memory (such as memorizing a paragraph of text).
- Women are more likely to suffer depression.
- Men are more likely to suffer autism.

Trying to pin these differences on innate, physical brain differences is roughly as difficult as creating cold fusion in a coffee cup. With the possible exception of the last item on the list, all of these differences can easily be explained by invoking the same force that makes us trim our hair, don designer jeans, and stop at traffic lights—the all-pervasive rules of *society*.

Gender Myths

The debate about gender differences has been clouded by more than few highly imaginative (and dismayingly popular) books written by people who should know better. Often, these books are criticized in scientific journals, but not before their bizarre claims leak into the popular media.

One of the most widely ridiculed claims argued that women speak 20,000 words a day, while men grunt their way to a mere 7,000. The source of the statistic was never found, and when scientific studies looked into the matter more closely they found that the daily word tally is nearly even—and in fact the men edge out the ladies. However, the original statistic lives on, because it agrees so well with what people expect.

 Feminist thinkers have argued that the myth of talkative women stems from the fact that many men would be happier if women weren't talking at all (possibly to avoid the inevitable fallout when female friends compare notes). After all, if you believe a woman should stay silent, even a normal amount of speech seems excessive.

A similar study thoroughly debunked the widespread belief that women have a phenomenal ability to gossip. Instead, the study surprised everyone when it found that men gossip at least as much as women and about mostly the same subjects, although they tend to place a slightly more overt focus on themselves.

Pink for Boys, Blue for Girls

The human brain is notoriously poor at separating the innate from the environmental. If we grow up seeing things a certain way, we come to expect that this way is natural, necessary, and unalterable. For example, if you've never seen a man in a dress, the idea seems absurd, and you should definitely not travel to Scotland.

A perfect example of this bias are color associations—the unwritten rules that make pink feminine and blue masculine. Although we accept these associations (even when we deliberately rebel against them), they're actually a relatively recent invention. In the 1920s, when Western parents began to dress their children in color, pink was preferred for boys. It was seen as a watered down version of red, and red was considered to be both masculine and fierce. Pale blue was used for girls and considered to be a far daintier color, possibly because of its association with the Virgin Mary. Somewhere in the 1940s, these preferences flip-flopped to the modern standards, where they've remained ever since.

On its own, this isn't too surprising. After all, the slickest salesman from a late night infomercial couldn't come up with a passable explanation for why men and women should have hard-wired color preferences engraved in their brains. But what's interesting is that this seemingly arbitrary detail is also a solidly entrenched and unquestioned gender stereotype—one that even the most progressive parents rarely challenge when dressing their newborns. In other words, color associations are one more reminder that the effect of our expectations can outweigh the reality in our genes.

As you learned in Chapter 7, your brain has an incredible ability to warp reality to make it fit your preconceptions. The link between women and talkativeness is one example. Here are a few more effects that can confound any attempt to figure out what behaviors belong with a specific sex:

- **Grouping bias.** Once we create groups, we exaggerate the differences between them and the similarities within them. The stakes are even higher when we belong to one camp. In fact, studies show that people behave differently in tests of aggression when they're told that their gender won't be identified. Men take advantage of the opportunity to kick back and be less assertive, while women exploit the freedom of anonymity to be more aggressive.

 Grouping bias may also be one of the reasons that correlations between behavior and gender fluctuate over one's lifetime. At different times of life, people may feel a stronger need to fit in.

- **Selection bias.** When we weigh the evidence, we only consider the facts we've noticed and remembered. For example, if you feel that women are genetically predestined to be poor drivers, you'll be quick to spot every erratic female driver on the road. Drivers who don't cause any problem will slip past your notice. Poor male drivers will be explained based on different rules and categorizations—for example, perhaps you'll classify them as arrogant teenagers, doddering seniors, or distracted yuppies.

- **Self reporting.** Even when they act in similar ways and pursue similar goals, the unwritten rules of society may lead women and men to describe themselves and their actions differently.

- **Power plays.** Everyone plays gender differences to their advantage, whether they're true or not. (For example: "I'm just a man. Don't expect me to stop watching this steel cage deathmatch and talk about emotions" or "I'm just a woman. My emotions overtook me, and I just *had* to buy the cutest pair of $350 shoes with your credit card. You're not mad, are you?")

Studying Young Children

There are two lines of evidence that suggest gender differences are more than just a collection of traditional myths, societal influences, and arbitrary biases. The first is the study of young children. The goal is to find behavioral differences that appear despite parental attempts to suppress them (such as aggression) or appear very early and so are unlikely to be caused by socialization (such as the face gazing of an infant girl).

Of course, it's rarely that simple. Young children are excellent groupers, and even before they fully understand the concept of gender they model their behavior after the gender they belong to. Slightly older children are quick to make conclusions based on sex (for example, they'll predict that a girl peer will choose to play with dolls, even if they're told she prefers trucks). And even infants younger than 12 months, who have no understanding of their gender, are treated differently by the adults around them.

With years of research to back up both sides of the debate, it seems unlikely that male and female behavior can be explained solely in terms of physical differences or social influences. Rather than try to convince you one way or the other, the following chart stacks up some compelling findings on both sides.

	Argument for Innate Differences	Argument for Social Conditioning
Communication	At birth, girls look at faces longer. By 12 months, girls make more eye contact than boys.	Mothers interact differently with babies depending on their sex. For example, mothers talk more to baby girls than to baby boys.
Sex-Typed Play	Boys tend to be more interested in vehicles (cars, trucks, and planes), weaponry (guns and swords), and building blocks. Girls favor dolls and role playing.	Children who learn to label gender earlier show more gender-specific play. Furthermore, when experimenters attach a specific gender to neutral toys (such as balloons and xylophones), boys and girls choose the ones that have been assigned to their sex.
Aggression	Boys are more likely to play aggressive games, such as ramming cars together. Girls show indirect aggression—they're sneakier.	Parents dissuade aggression in boys and girls, but have a higher threshold with boys. As infants, girls are handled more gingerly by their parents and boys are stimulated with rougher physical play.
Empathy	Baby girls respond more empathically to the distress of other people, for example by displaying sad facial expressions when they hear another infant crying.	Similar reactions are often characterized differently based on the gender of a baby. In one study the surprise of a jack–in-the-box was most commonly described as anger if onlookers were told the infant was a boy, or fear if they were told it was a girl.
Social Play	When joining a group of peers, girls are more likely to watch and wait. Boys are more likely to interrupt. Girls are also more willing to recognize a newcomer, while boys are apt to ignore others.	When observing behavior, children are more likely to follow the example set by a person of the same sex.

Studying Wrongly Sexed People

Some of the most compelling examples of innate sex-specific behavior appear when you consider the most unusual cases—for example, instances where the ordinary pairing of gender and sex hormones is disturbed.

In 1965, a botched circumcision with an electrocautery needle burned away David Reimer's entire penis. According to the wisdom of the time, he was reassigned as a female. Doctors removed his testicles, and fashioned female genitalia. His parents changed his name to Brenda, and he was raised as a girl. When Brenda reached adolescence, estrogen supplements were used to trigger puberty, causing her to develop breasts. For the leading-edge doctor who oversaw the case, Brenda was the ultimate experiment to prove the power of social conditioning over raw biology.

Except it didn't work. At two years old, Brenda tore off dresses and fought to get toy cars and guns from her twin brother. In school, she was bullied mercilessly for her apparently masculine traits. At home, she told her parents that she felt like a boy. When she finally learned the truth at age 14, Brenda felt a sense of deep relief. With the help of surgery, Brenda embarked on a new life as a man named David. (Sadly, David never fully recovered from his ordeal, and he committed suicide at 38.)

David's story offers strong evidence that some details of gender identity are fixed before birth, as they are in other animals. Presumably, David's development in the womb was fairly normal. The SRY gene on his Y chromosome kicked off the usual process of male-ification, and testosterone passed into the brain where it made deep, irreversible changes. And even though his development as a man was interrupted after birth by the removal of his testicles, these changes prevented him from becoming a well-adjusted woman.

This is one of the most dramatic examples of sex reassignment gone wrong. However, there are other cases that hint at the powerful effect of testosterone in the fetal world. One example is found in girls who suffer from a genetic condition called *congenital adrenal hyperplasia* (CAH). This condition causes them to produce testosterone in the womb. Depending on the amount, their female genitalia may become enlarged and more masculine. Studies of CAH girls have been conducted in half a dozen countries, and they consistently find that CAH girls act a lot like boys. They prefer to play with boys, they favor rough play, and when choosing toys they prefer cars, trucks, and guns. As adults, they're more likely to have homosexual relationships.

Note A similar problem happens to males who suffer from ***androgen insensitivity syndrome*** (AIS). Although they're genetically male, their bodies lack the hormone receptors that pay attention to testosterone. As a result, they develop as females (complete with female genitalia), and they describe themselves as female , without having any idea about their genetic reality.

The Statistical Picture

Agreeing that there are differences between men and women is only the start. In order to really understand what's happening, you need to know how the differences are ***distributed***.

For example, imagine you run a study that explores how children play with dolls. You find that on average, boys take longer to pick up a doll than girls, and play with it for shorter amounts of time. When you describe this result to your friends, they probably imagine that the results are something like this:

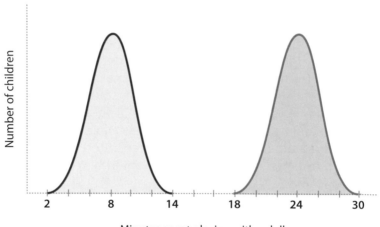

Minutes spent playing with a doll

Here, the blue curve shows that boys spend much less time playing with dolls. The average play time for boys is 8 minutes, but some hold on to it for just 2 minutes while others keep it for 14. Girls range from 18 to 30 minutes of doll play, with an average of 24 minutes.

The Riddle of Homosexuality

If talking about politics, religion, and gender differences isn't enough to offend most of the guests at your next dinner party, here's another controversial topic to bring up: homosexuality. After all, you can't spend much time looking at the biological processes that create boys and girls without wondering if gayness is a biological imperative or a personal choice.

Numerous studies have found differences between gay and straight men and women, including subtle shifts in brain anatomy (particularly in the hypothalamus, which plays a vital role in sex behavior). However, these differences don't indicate *causality*—in other words; they can't answer whether the brain triggers homosexuality or homosexuality reshapes the brain. After all, the brain's structure changes with use. Saying that homosexual people have different brains might be like saying that athletes have bigger muscles—in this case, a genetic component might be at work, but the real difference is the lifestyle.

Other studies search for differences in biological details that are set much earlier in life and aren't likely to change based on a person's environment. For example, various studies suggest there are differences in the handedness (the preferred use of the right or left hand), fingerprint patterns, finger length, hearing, and armpit secretions in gay people. Furthermore, twin studies (a type of study that's described on page 241) suggest that there's a genetic component at work—in other words, gayness runs in families. Lastly, studies of other species tell us that it's not just genes but possibly the prenatal environment that casts the deciding vote. If experimenters change the mix of hormones in the fetal world, the sexual behavior of that animal is changed forever.

So where does all this leave us? In much the same place that we're at in the faceoff between masculinity and femininity, with good reason to suspect there are biological processes at work, no way to be sure, and a responsibility to tread cautiously. Here are some final points to keep in mind:

- Research has found that gay adults usually show gender nonconformity as young children. This suggests that no matter what the cause, the path to same-sex attraction starts with changes that take place early in life, before puberty. In other words, the roots of sexual orientation don't lie in tofu overeating, immoral television, or the whims of college experimentation.

- Sexual orientation develops across a person's lifetime. Different people realize at different points in their lives that they are heterosexual, homosexual, or bisexual.

- There is no scientific evidence that sexual orientation can be changed at will—say, by the simple act of therapy.

If this graph is accurate, your doll study has turned up some highly unusual results. Most of the time, the differences between the sexes have a generous amount of overlap. Even if girls, on average, clock more doll time, the full set of results is more likely to look like this:

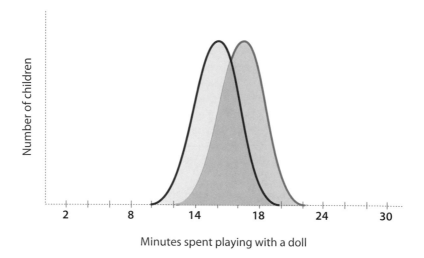

Minutes spent playing with a doll

Here, the same conclusion applies—namely, girls spend more time on average with their dolls. However, whereas the first result showed a dramatic difference between all boys and all girls, this more realistic study shows that most boys and girls will fall somewhere in the broad middle, and play with the doll between 12 and 20 minutes.

> The distinction between these two graphs shapes how we understand sex differences. It's particularly important because when we discover differences between sexes, they all too easily become part of our definition of what it means to be a man or a woman. For example, if a researcher finds that men excel at math (in other words, have a slightly higher overall average), it becomes easy for teenage girls to give up on the subject and parents to excuse poor performance.

In these two examples, the shape of the distribution stays the same for boys and girls. However, it's also possible for certain characteristics to vary more among one sex than the other, which changes the pattern more dramatically.

This sort of theory is sometimes applied to suggest that men have more *variability* in psychiatric illness and intelligence. In other words, the average man and average woman stack up equally well on an IQ test, but there are more outliers—exceptionally good or exceptionally poor cases—in men than in women. Here's a chart that shows this sort of distribution:

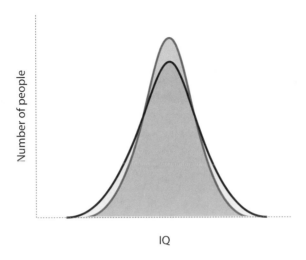

Theories like this are extremely controversial, because they suggest there's a good genetic reason why men and women dominate certain careers. For example, if theoretical physics is a discipline that only draws on the extreme outliers—the off-the-chart geniuses who fall at the extreme right of the graph—there won't be many women around the particle accelerator.

 Note Harvard president Lawrence Summers mused about this effect being one possibility for the underrepresentation of women in technical fields like engineering (along with institutionalized discrimination and the conflicting demands of work and family). Shortly afterwards, he was forced to resign.

Love and Relationships

As you've seen, the quirks of biology split men and women into two opposing camps. Fortunately, biology uses its best trick to bring the two sides back together.

Coming to Terms with Sex Differences

If you were hoping to end this chapter with a perfect explanation of why women act like women and men act like men, it's time to abandon your quest. As you've learned, genes, hormones, and civilization are locked together in a complex relationship that's as tangled as a bowl of overcooked spaghetti. However, the search for sex differences teaches some important lessons:

- **Don't assume.** Many pieces of the gender picture are missing or out of place. In the meantime, it's safe not to assume too much about what characteristics are necessarily female and what ones are unavoidably male. After all, less than 100 years ago the nearly universal male opinion was that women were too wholesome, frail, and emotional to be allowed anywhere near a polling booth.

- **Don't get too cocky.** If you're a man, you might be tempted to brag about the high-variability theory of intelligence. But unless you're one of the extremely rare outliers on the correct end of the scale, there's no reason to rejoice.

That trick is *love*, the brain-melting phenomenon that's behind a range of borderline psychotic behavior (such as writing tortured erotic poetry) and actual psychotic behavior (such as jumping out of a cake in a half-naked cupid suit to deliver that poetry). And though you might assume that love is just the work of pumped-up lust, it turns out that it's quite a dramatically different thing.

The Love Effect

Examining the brain in love is relatively new ground for neuroscience. However, a few love studies provide tempting clues.

In 1999, a study concluded that falling in love is rather like acquiring a psychiatric illness. The study tested a group of college students who had recently fallen in love, but hadn't consummated their new passion. These new lovers had significantly lower than normal levels of serotonin, a key neurotransmitter that plays a host of subtle roles in the brain. In fact, their sagging serotonin levels were in line with sufferers of OCD (*obsessive compulsive disorder*).

People with OCD suffer from fixations, recurring thoughts, and compulsions that urge them to perform repetitive rituals (like washing their hands, locking doors, and counting objects). There's no definitive explanation for why serotonin is low in OCD patients—it could be a cause of OCD, an effect, or a more subtle interplay of factors. However, it takes but a small imaginative leap to see the similarity between OCD symptoms and the single-minded infatuation of new lovers.

Several more recent studies have looked at the brain activity of new lovers as they gaze at a picture of their beloved. These studies find that the brain's reward system (namely, the nucleus accumbens that you learned about on page 128) lights up like a fairground when it recognizes its loved one. The effect is pronounced—in fact, the human brain juiced-up on love more closely resembles the euphoria and cravings of a junkie than an ordinary person ogling a pornographic magazine.

These brain imaging studies suggest that love is a powerful biological drive that extends far beyond the basic urge to merge. This goes a long way to explaining the strength of its effects. While lust can bring people together for an afternoon, only love is able to wed them together for several years or more.

The Timeline of Love

So far, the studies you've learned about deal with people in the early throes of love. Follow-up studies show how the activity changes in the brains of lovers as time wears on, and the conclusions are remarkably consistent.

Essentially, love begins with a blazing period of infatuation. This is the state that volunteers were in when the studies concluded they might as well be half-mad or hopped up on illegal substances. This stage overpowers reason, makes us feel good, and helps the brain build strong addict-like associations between imagined pleasure and our loved ones.

However, it doesn't last. In both studies, the brain was back on an even keel a year later, with normal serotonin levels and less activity in the brain's pleasure circuit. That doesn't mean the early love had died out, only that it had shifted into an altogether different experience—a long-term bond of mutual dependency and companionship.

From an evolutionary point of view, the sequence of events seems almost too perfect:

❶ First, the brain's hard-wired reward system creates the cravings we call lust. This force urges people to look for mates. Without it, we'd be more likely to stay in our caves watching reality television until the human race fizzles out.

❷ When the brain finds a suitable mate (and no neuroscientist has a good theory for how this selection process works), the heady passion of love sets in. This wild reaction is often enough to get us over the real hurdle—settling down to raise the fruit of our loins.

❸ The final stage is the least certain. It seems likely that humans are designed to be monogamous—to a point. Unlike most other animals, human babies spend an embarrassing amount of time being utterly adorable and utterly defenseless. For that reason, it makes sense that evolution would favor parents who stick around long enough to protect their children and teach them how to survive. That said, there are also evolutionary incentives to cheat stealthily while in a committed relationship, or hook up with someone younger after your children hit puberty. Ultimately, it's up to you whether you believe that love has an inevitable expiration date.

Bonding

As you've seen, love is a genetic program that begins with white hot passion and sexual bliss, and ends up knee deep in soiled diapers. The brain needs to work hard to keep a couple together through this transition. The pleasurable associations that we build up around our loved one in the infatuation stage certainly help. However, they probably aren't enough to last a decade of child rearing. Fortunately, there's another ingredient that comes into play: the somewhat mysterious compound known as *oxytocin*.

Oxytocin is an unusual hormone that has effects both in the body and in the brain. In the female body, it stimulates the uterine contractions of labor and the letdown reflex that allows breastfeeding. In the brain, it acts as a neurotransmitter. The hypothalamus releases it during hugging, touching, cuddling, and orgasm in both men and women.

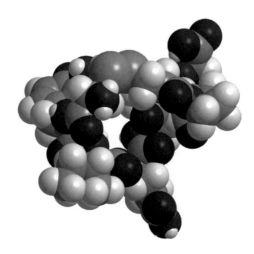

The effects of oxytocin are controversial, but solid studies suggest it plays a key role in trust, maternal behavior, and emotional bonding between family members and between sexual partners. Here are some of their findings:

- **Oxytocin seems to promote pair bonding in prairie voles.** Prairie voles are legendary for their monogamous relationships, which are highly unusual in the animal kingdom. However, if oxytocin levels are disrupted, the formerly faithful voles start to wander. (Incidentally, prairie voles pick their partners after getting a whiff of irresistibly decadent vole urine.)

- **Oxytocin seems to promote maternal behavior in other animals.** For example, when scientists mess with the oxytocin levels in rats or sheep, they're less likely to take care of new offspring. And in one study, spinal oxytocin injections made virgin sheep discover their mothering side.

- **Oxytocin seems to promote trust in humans.** In one of the most curious oxytocin studies yet, volunteers playing an investment game were more likely to trust a stranger after they'd inhaled a nasal spray with oxytocin in it.

- **Oxytocin might keep the pleasure circuit running.** As you learned in Chapter 6, the brains of habitual drug users adapt to ratchet down the pleasure of their drug, turning a formerly ecstatic experience into little more than a brief moment of relief. However, in studies with rats, oxytocin reduced this tolerance effect. That means that oxytocin might help the heady passion of new love remain effective for a longer period of time than any chemical hit.

- **Oxytocin might fuel the illicit drug MDMA (known as Ecstasy).** MDMA has several effects in the brain. It prevents neurons from cleaning up leftover neurotransmitters, which keeps the brain's pleasure circuit active. However, MDMA also promotes the release of oxytocin, which might be the cause of the "loved up" feeling users describe. (If you're thinking of trying it out for yourself, see page 133 for a few reasons why tampering with the brain's pleasure circuit isn't such a great idea.)

However, no one knows exactly what oxytocin does. Some argue that it dampens down negative emotions, like the fear response from the amygdala, which could otherwise interfere with bonding. Others suggest that oxytocin smoothes out the highs and lows of the brain's pleasure circuit, making it easier for people to transition from the highs and lows of sexual orgasm to the lasting bond of a romantic partnership. Oxytocin receptors are found in the neurons of both these areas of the brain, and in other areas as well.

However, a few other studies suggest that oxytocin isn't quite so simple. For example, women in stressful relationships have higher than normal levels of oxytocin. In this case, it's possible that the brain is ramping up its oxytocin producing abilities to attempt to gloss over the problem or prompt the sufferer to get out and find a new mate. Either way, it's clear that oxytocin doesn't generate cuddly apple pie feelings on its own. Instead, oxytocin is part of a brain mechanism that creates bonds. Neuroscientists have yet to determine whether it's the most important part, or just one component of a complex biological process.

The Habits of Highly Successful Relationships

Surprise! Your understanding of neuroscience can help separate solid relationships from short-term flings.

The key thing to remember is that love, like hunger and thirst, is a biological drive. You may believe that there's only one perfect match for you, but your brain certainly doesn't agree. In fact, as soon as you settle on someone, it pulls out a hefty bag of biochemical tricks to move your relationship along. That's why people are remarkably successful at finding their soulmates, whether they hook up in clubs, on Internet dating sites, or through the formality of an arranged marriage.

Here are some neurological tips for successful romance:

- **Don't expect to perceive a loved one uncritically.** Critical thinking diminishes in a brain that's in love. So if you *feel* a partner's completely right for you even while you *think* they're bad news, don't expect the happiness to last. The feelings will fade while the reality remains.

- **Remind yourself of the relationship timeline.** For most people, the unrestrained passion of early love mellows in a year. That fact might mean a few less giddy smiles and vacant stares, but it's nothing to be worried about—after all, nonstop no-holds-barred passion can be exhausting. Successful relationships adapt to these changes to set down a solid partnership for the future. Less dedicated partners pack up their bags and go on in search of the next love fix.

- **Don't pin everything on romantic love.** It's fun while it lasts, but it isn't the best predictor of a lifetime of happiness. Romance, sex, and passion change—and if they aren't allowed to evolve, they'll simply sputter out.

- **Remember that relationships are work.** People prefer to enjoy the pleasure of love and let their brains figure out the rest. This approach is enough to create a relationship, but not necessarily a good one. Instead, while your brain automatically takes care of pleasure and bonding, you need to put in the heavy lifting—and that includes everything from staying faithful to having difficult discussions about problems and feelings.

The Practical Side of Brain Science

Selecting a Mate

Falling in love is the easy part. If you want to build a relationship that lasts long after your brain's pleasure circuit powers down, you need to take a closer look at your partner. For long-term relationships, you need to tick the following boxes, all of which suggest that a lover is a good long-term match:

- ☐ Does your romantic partner give you support, affection, and respect?
- ☐ Does it feel effortless to be together?
- ☐ Do you like yourself when you're with this person?
- ☐ Do you have a sense of shared purpose?

Researchers find that even in the initial stages of a romance, it's surprisingly easy to predict which relationships will last. One of the best tests is to watch a couple argue. Couples who last have a much more effective negotiating style, and the ratio of positive to negative statements during a conflict conversation is five to one. Relationships with grimmer prospects have more negativity, and a positive-to-negative ratio that's closer to one to one.

Just as clear are the warning signs, which psychologist John Gottman calls the four horsemen of the apocalypse. They include contempt, direct insults, sarcasm, and an unchecked feeling of superiority. Nearly as bad are criticism, defensiveness, and emotional withdrawal. If you hit one of these bumps, fix it fast, before it corrodes the foundation of your relationship.

10 The Developing Brain

You can't spend 222 pages exploring the quirks of your brain without starting to wonder exactly how it was created. Of course, you know some of the circumstantial details—the man, woman, lingerie part of the equation—but that hardly accounts for the 100 billion tiny electrical links that create love, consciousness, and run on sentences. To get the full picture, you need to go back to the very first hour of your life, and then you need to go back a bit further. In fact, you need to start at the very beginning of your beginning, the icky moment when your parents got together, did unspeakable things, and set your brain on an irreversible course from a single cell to reading this book.

In this chapter, you'll take that journey and see how your brain developed from conception, passed through the rocky waters of teenage life, and ended up in the middle of adulthood (if indeed you've made it that far). You'll see how biological processes work like a sculptor to chisel away nearly half your neurons before you've had a chance to use them. Finally, you'll look forward and consider how your brain changes as you glide into the twilight years of old age.

Before Birth

You know the script. Tragically outdated clothes, a sordid night of passion, and moments later a single fertilized cell was on its way to becoming you.

Early on in your development, when you were little more than a teaspoon of jelly, your brain began to form. It started out as a disk of rather unremarkable cells that appeared about two weeks after your conception. Over the following week, a groove appeared in the middle of this plate, and by week three that groove wrapped itself into a closed cylinder called the *neural tube*.

Neural tube

| Neural groove forms | Neural groove closes | Neural tube forms |

The neural tube is the place where the entire nervous system is built. In humans the neural tube develops into a spinal cord topped by a brain, which bulges up over the following weeks like a hastily inflated party balloon. At the seven month mark the brain begins to develop the deep bumps and folds called *gyri* and *sulci*, which give your brain more room for heavy thinking.

The prebirth process of brain building is staggeringly complex:

- In a relatively short period of time, your brain produces the billions of neurons it needs for a lifetime of thinking.

- Each neuron has to crawl through the neural tube to the right location. The brain builds itself from the inside out, and the outermost neurons of the brain—the deep thinking layers of the cerebral cortex—need to push their way from the innermost part of the neural tube and through a mass of cells to get to their genetically determined positions.

- Each neuron has to develop into a specialized type that's appropriate for the role it's going to play. For example, the neurons that control muscle movements (motor neurons) are different from the neurons that detect light and the ones that respond to pain.

- Your neurons begin to grow the axon and the dendrites (see page 16) that will link them to other neurons.

- At the same time that it's frantically building neurons, the brain also needs to stock itself with billions of glial cells. These are support cells that perform a variety of housekeeping tasks in a mature brain (like improving signal speed and cleaning up debris). They also help to guide the development of a new brain.

At the end of this process, the brain gets ready to do something wholly unexpected—it prepares to kill off billions of its own neurons through a process called *apoptosis*. If you're used to associating dying neurons with doddering old age, this will come as something of a surprise. But from the body's point of view, it makes perfect sense. By the seven month mark, the brain has overbuilt itself—in fact, it has twice the neurons it needs. To weed out the excess, a fierce competition begins. Neurons struggle to bind to other neurons and drink the nutrients they need to fuel their development. The weaker ones shrivel away, leaving a fitter brain behind.

At birth, the brain has about 100 billion neurons—and with a few minor exceptions, these are all the neurons you'll have for the rest of your life.

 Note There's good reason to expect that even at a mere 100 billion neurons, the brain has all the mental hardware it needs for several lifetimes' worth of learning. More important is the number and strength of the connections that link your neurons together.

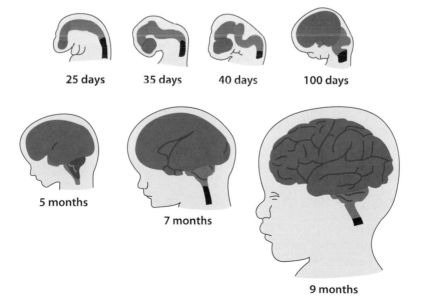

25 days   35 days   40 days   100 days

5 months

7 months

9 months

Parenting a Fetus

It today's world of high-stakes child rearing, parents worry about every stage of the process—including what takes place inside the womb. Now that you've seen the astounding sequence of events that turns a fluid-filled cylinder into a complete nervous system, it's time to put those concerns in perspective with a few tips:

- **Don't worry.** This is one of the few points in your child's life where you won't need to worry where they are, what they're doing, and how a two-year old can eat an entire box of crayons in the time it takes you to flush the toilet. If you feel inclined to talk to your abdomen and play classical music, by all means do. But don't worry if you'd rather let nature take care of this part, because it's unlikely that the developing brain needs any extra excitement.

- **Take folic acid.** Folic acid can significantly reduce the risk of various neural tube defects. But there's a catch—you need to start well before you conceive. Some doctors suggest that all women of childbearing age take prenatal vitamins.

- **Avoid alcohol, drugs, and mysterious herbal teas.** The time before birth is extremely critical, because the brain needs to create the full set of neurons that will last through life. Although the brain of a developing fetus is resilient, some substances can disrupt the way neurons develop and move through the neural tube. The best known examples include alcohol, illicit drugs, and many prescription medications.

Childhood

Ah, childhood. A time of innocence, exploration, and massive synaptic loss.

Sound odd? As you've already seen, the brain kills off extra neurons before it enters the world. If the development of the brain was like making a house, the construction workers would build twice as many rooms as you need and then demolish half of them before letting you in the front door. A similar phenomenon happens with *synapses*, the connections that link neurons together. Through life, the brain strengthens the best connections and prunes away the weakest. However, this phenomenon is particularly pronounced at two points in your life—as a young child facing the world for the first time, and as an adolescent entering the teenage years.

Wiring the Brain

Wiring a brain is somewhat like sculpting a statue. You begin with more than enough stone (in the form of excess neurons before birth and excess synapses during childhood). The craft lies in chiseling away the excess until you're left with the form you want.

The figure shown here compares the connections between neurons from birth until two years. The number of neurons doesn't change. However, as the child develops, each neuron sends out a thicket of dendrites in search of other neurons. It's a bit like a lonely partygoer hunting for friends to talk to.

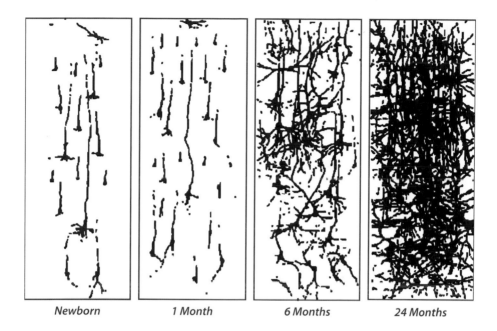

| Newborn | 1 Month | 6 Months | 24 Months |

After 24 months, this wave of synaptic growth reaches its peak, leaving a heavily-connected brain and an emotionally unpredictable two-year old. This is when *synaptic pruning* ramps up. Frequently used connections strengthen, while neglected ones gradually shrivel away. This is one of the reasons that newborn babies can distinguish between more speech sounds than adults, teenagers, and even one-year-olds. As babies master a language, they stop paying attention to the sounds that aren't important, and those connections are trimmed away.

Estimates suggest that the baby brain loses as many as 100,000 synapses each second at the height of its development. As an adult, your brain retains little more than half of the synapses you had as a two-year-old.

Incidentally, a similar pruning process happens with other animals, but on a lesser scale. Rats prune out just 10 percent of the connections in their cortex, while cats lose 30 percent. The difference in humans is usually attributed to the complexity of our brain—essentially, it's more difficult for neurons to create precise connections through the tangled undergrowth of a human brain. It's also possible that greater synaptic pruning is a process that helps make humans so remarkably adaptable to different environments.

Synaptic pruning is one way that the brain reshapes itself into a lean thinking machine. Another important process is *myelination*, where ordinary neurons are wrapped in a sheath of insulating fat. Myelination is important because it lets signals travel along a neuron faster and with less degradation. It's the difference between setting up a home theater system with top-quality cables and hooking it up with paperclips and elastic bands.

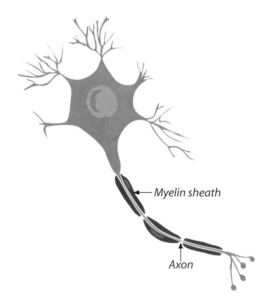

Myelin sheath

Axon

Interestingly, myelination appears to take place in a preset genetic pattern that's impervious to high-powered parenting. In other words, you can't teach a 4-month-old to walk. Instead, when the motor neurons that control leg movements are myelinated around the one-year mark, the baby is ready to stand up and take a first step.

Note As neurons are insulated, babies change—often abruptly. If you're a parent who worries about developmental milestones, take a deep breath and remember myelination. When the correct brain hardware is in place, a baby will master new skills with alarming speed, and the infant who can't roll over one day is the same one you'll find crawling through your collection of rare petunias next week.

The myelination of the brain is also the key reason for its dramatic growth. At birth, the baby brain weighs in at a modest one pound—less than the size of many diet-busting hamburgers. Within the first year, the brain doubles in weight and size. By age 5, the brain has picked up a full two pounds, and has reached 95% of its full adult size. That growth from one pound to three is the result of longer dendrites, new glial cells, and myelin.

Critical Periods

The brain is programmed to connect neurons, prune synapses, and insulate axons at specific times in childhood. In a few cases, this leads to a *critical period*: a narrow interval of time in which a specific skill has to be acquired in order for development to be normal.

The most famous example of critical periods is found with vision. If a cat lives the first three months of its life with a blindfold over a perfectly normal eye, it will lose the sight in that eye forever. The same experiment has no effect on an adult cat—its fully developed sense of vision, if not its dignity, holds up to months of blindfolding.

This difference occurs because the developing brain is a highly competitive environment. Neurons that are doing something useful—such as receiving information from a fully working, uncovered eye—will win the battle for resources and take over more brain real estate. (Oddly enough, scientists have shown that this phenomenon can happen the other way around. For example, if a third eye is grafted onto a frog embryo, that third eye will fight for neurons and the odd-looking frog will develop three perfectly wired, working eyes. But no one knows exactly what the frog makes of the experience.)

 Note The brain's ability to reshape itself is called *plasticity*, and it's most apparent in a young, developing brain. That's why children can suffer severe brain damage and still show surprisingly normal development, while the same damage in adulthood is likely to prove catastrophic.

Humans have similar critical periods for visual development, certain motor skills, and language, although in this case the window is much longer. Children have several years to learn a language, but if they don't pick it up before adolescence, language skills like pronunciation and grammar will be irreversibly affected.

 Note The critical period for language learning begins to close around 5 years of age. If you want your child to speak a foreign language like a native, introduce it before age seven, and make sure your child keeps speaking it.

The idea that critical periods might apply to every type of learning has caused a great deal of panic in otherwise sensible parents. However, the reality is that children continue to develop and whittle away synapses throughout their entire lives. If your child misses the chance to learn an instrument, sport, or craft, don't worry—they'll have ample opportunity to master it through adolescence.

Baby Myths

New parents are easy prey for calculating marketers. They haven't been bled dry yet by tuition, orthodontics, and teen fashion. They're desperately motivated to create the best environment for their baby. And they're too sleep-deprived to tell the difference between a bottle of vinegar and a carton of milk.

If you find yourself in this situation, consider these recommendations. First, watch for vinegar in your cereal. Second, look out for the following insidious myths:

- **Super-enriched environments create smarter children.** In an often-cited experiment, researchers found that rats in ordinary cages had smaller brains than rats that got to play with all sorts of whizzy toys. Since then, several scientists have pointed out that the enriched rat environment corresponds to a normal Western household, while the deprived environment—confined in a dull cage with little chance for exploration, exercise, or social interaction—would be considered abusive for a human child. The balance of research now shows that a severely deprived environment will harm a child's development, but the difference between an enriched environment and an even more enriched environment is vanishingly small.

- **Learning toys promote development.** Research clearly identifies factors that harm the development of a baby brain, including poor nutrition, environmental toxins (like lead), drug exposure, and chronic stress. However, toys that are specially designed for intellectual stimulation have no apparent effect, despite their often imaginative claims. Baby Einstein, this means you.

- **Childhood is a race to acquire facts.** The most important childhood learning is skill-based (how to investigate the world and interact with other people) rather than fact-based (the names of different animals, colors, numbers, and so on). In other words, think twice before you sit a young child in front of an educational television show. They may pick up a few facts, but they'll lose valuable time that could be better spent interacting with their environment.

- **TV is a learning aid.** Educational DVDs are tremendously popular, partly because every parent could use a 15-minute interval to shower, answer a telephone message, or put out a raging stovetop fire. However, when it comes to TV watching, parents who focus on the content of the program may be missing the point. Some studies suggest that television watchers are likely to develop connections that track fast-moving objects while crowding out room for slower-paced exploration and social development. A recent study suggests a link between DVD watching and smaller vocabularies, presumably because time in front of the TV is time that isn't spent talking to other people. And while 49 percent of parents think educational DVDs are very important in the intellectual development of children, only six percent know that the American Academy of Pediatrics recommends children under two avoid it altogether.

If you're still feeling uncertain about your decision not to shell out for that $300 disco playcenter monstrosity, reassure yourself with the following baby truths:

- **Children are natural learners.** They're quite capable of seeking out the stimulation activities they need.

- **Variety is good—and easy.** Exposing children to a wide range of different experiences is one of the top jobs of parenting, and for young children it's easy. At a young age, a walk to the grocery story can be as educational as a trip around the museum.

- **Relationships are most important.** Studies show that when it comes to school readiness, how your children feel is more important than what they know. A child who can rely on nurturing, dependable relationships will gain the confidence to grow, explore, and ultimately ask for that $300 so she can buy an iPhone.

The Teenage Years

It's a time of turbulence, when hormones rage, tempers flare, and the brain's logical thinking systems go offline—and that's just the parents. Whether you're living through them or parenting someone who is, the teenage years have a well deserved reputation as a trying time.

In the past, scientists believed the teenage brain was essentially the same as the adult brain, minus a few life lessons. The infamous teenage moodiness was chalked up to the effect of the sex hormones you learned about in Chapter 9. However, several new studies have uncovered dramatic evidence that the teenage brain is still a work in progress.

Here are some of the events that happen to the teenage brain:

- **A second wave of synapses grow.** Between 7 and 11 years, the brain repeats the same trick it used in the first two years of its life. It produces a huge growth of dendrites that stretch out in search of other neurons. This second wave happens just before puberty, but it's not linked to it—for example, if puberty is delayed for other reasons (such as poor nutrition), this brain boost still takes place. As a child becomes a teenager, the synaptic pruning begins again.

- **Myelination continues.** The myelination process that began in childhood is still underway. The areas that are myelinated last include the prefrontal cortex (page 148), which forms the seat of higher reasoning and impulse control. It isn't fully online until the age of 18 to 20.

- **Patterns of brain activity are different.** When showed pictures of faces with emotional expressions, adults use the frontal regions of their brain to identify them. When teenagers look at the same expressions, they use the *amygdala*, the tiny brain area that governs instinctive emotional responses like fear (page 134). This difference suggests that teenagers are more likely to respond to other people with an instinctual, emotional reaction. Even more interesting is the finding that adults had no trouble identifying emotions in the facial expression test (fear), while teenagers consistently thought up similar but slightly off-the-mark interpretations (surprise, shock, anger). This suggests that teenagers might have a sound neurological excuse for misinterpreting parents.

- **The cerebellum changes.** The *cerebellum* is the odd growth on the back of the brainstem that's a bit of a mystery. It plays a role in coordinating movement, but recent research suggests it plays a subtler role coordinating different activities in the brain.

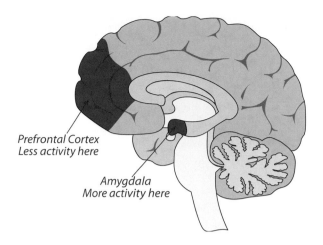

Prefrontal Cortex
Less activity here

Amygdala
More activity here

The fact that the teenage brain is still developing can explain a lot of emotionally charged, unpredictable behavior. However there's also another way to look at the tribulations of teenagedom—the evolutionary perspective.

In the world of our distant ancestors, life expectancy was dramatically shorter. Young adults didn't have decades of life to kick around in their parents' basement playing *Gears of War*, because their parents might not make it through their 30s. From the brain's point of view, adolescence was a do or die moment—a time to ramp up development and prepare for the challenges of adulthood.

Teenagers who were a bit more relaxed about life, and didn't feel the same emotionally-charged hormone-fuelled urgency to go out and create the next generation might succumb to disease or have a bad run in with an ill-tempered hyena before getting the same chance again. In fact, it sounds like a parent's worst nightmare—the easy-going teenagers died out while the sex-crazed early developers inherited the earth.

 Biologists tell us that most other animals—including the great apes—have the good sense to skip the teenage transition phenomenon and glide easily from infancy to adulthood.

An Owner's Guide to the Teenage Brain

If you're the lucky owner of a teenage brain, you're in a rare period of life where your potential dwarfs your past. It's a time of opening vistas, expanding horizons, and endless social embarrassment. Here are three tips to make the turbulence easier to manage:

- **Train your brain.** The second wave of synapse growth provides a great opportunity for learning new skills. Now's the time to pick up a ukulele, write a novel, or learn to breakdance. You can always pick up these interests later in life, but you'll excel more easily if you lay the groundwork now.

- **Get a decent night's sleep.** The teenage brain can't get by with feeble adult-sized portions of sleep. Instead of 7 or 8 hours, you need 9 or 10. And to make matters worse, the circadian rhythm (page 46) of your teenage brain shifts, so that it expects to stay up later and sleep-in longer. Unless you can convince your local school to start class later (and some school districts are doing exactly that), you'll need to compensate by going to bed earlier than you want to.

- **Avoid brain-disrupting chemicals.** It's an unfortunate coincidence that the time when your brain is most interested in expanding its horizons and taking new risks is also the moment it's most vulnerable. Studies have consistently found that while binge drinking, nicotine, and illicit drugs like ecstasy are bad news for adults, the effect is magnified in developing brains. The decisions you make to scramble your brain now could leave it off kilter for the rest of your life.

Parenting a Teenage Brain

Now that you've seen the upheaval of the adolescent brain, you probably have a bit more sympathy for the temperamental, drama-generating, texting, cellphone-addicted being who lives in your house. And while this book would need to borrow pages from *War and Peace* to describe the real art and pain of teenage parenting, there are a few simple things that neuroscience tells every parent:

- **Expect emotional outbursts.** In a teenager, volatile moods aren't the result of a defective brain, but a response to a constellation of new pressures. In a relatively short time, teenagers are driven to distinguish themselves, experience life, and participate in the social dynamics of a peer group.

- **Don't try to win arguments.** It doesn't take many words to state your case and convey your unwillingness to bargain. If you're drawn into a senseless argument, you will lose. Teenagers rapidly master the art of explosive negativity, and they'll enlist every logical fallacy from Chapter 7 to prove that life isn't fair and "You're not the boss of me." Remember, your brain is myelinated, your synapses are neatly pruned, and you don't have an excuse when your emotions creep into the fray.

- **Incorporate teens in the real world.** Some suggest that the slightly absurd, alternate universe of teen culture exists because teenagers aren't yet accepted as full members of society. Cultures that erect fewer barriers between the adult world and the adolescent world often seem to navigate teen troubles more easily. For example, societies with a low drinking age (or none at all) generally have teenagers who are more responsible drinkers. Similarly, European cultures that give teenagers more freedom generally have an easier time integrating them into society.

- **Let teenagers have a safe environment to express risky impulses.** In other words, favor reckless skateboard tricks over reckless drag racing. Overprotective parents probably don't create risk-taking kids (that's more likely to be a genetic personality trait). However, they do risk raising kids who lack something important—namely, the basic framework of experience that helps teenagers tell the difference between activities that will end with skinned knees and those that lead to broken bones.

- **Keep talking.** Yes, it will embarrass the heck out of them. But they'll still listen. And one of the most practiced skills in every teenage arsenal is pretending not to listen (followed closely by pretending not to care).

Old Age

Your brain develops at a breakneck speed throughout your early years. From an evolutionary point of view, this makes perfect sense—after all, our ancestors needed every advantage they could get to survive the harsh and brutal prehistoric world long enough to have babies. Unfortunately, once you've passed your genes along you've ensured your evolutionary success, and your personal survival is decidedly less important.

Although we're used to associating old age with the very last decades of life, you may be closer to it than you think. If you've passed the age of 20, your brain has begun its long and steady decline. Your family of neurons, which you've held since birth, is beginning to show some serious wear and tear. As each year passes, your brain shrinks a bit more.

Here's how your brain changes as it ages past 20:

- **The brain shrinks.** Brain size peaks at 20, and if you reach 100 you may have made it there with 15% less brain. The actual cause for the shrinkage is controversial. Some suggest it's neuron loss, others point to the breakdown of myelin around neurons, while others think it's the result of continued synaptic pruning.

- **The brain slows.** As we age, our reaction times slow down. When given problems, we reason more slowly and take more time to assemble a plan. Already at 30, neuroscientists can measure performance differences between our current performance and our better 20-year-old selves. Recall is slower and information lingers in short-term memory for a little less time.

- **Memories fade.** Memory is one of the best known failings of advancing age. And it doesn't just apply to ancient events. The older our brain is, the more difficulty we'll have using associations to stitch together the details of recent experiences as well.

- **IQ and language hold fast.** On average, old people perform just as well on most IQ and language tests. Most neuroscientists believe that this shows a tradeoff between efficiency and raw brain power. In other words, even as our neural hardware is starting to rust, we're becoming more experienced at using it and coaxing out every last bit of mental performance we can.

This list paints a grim picture, but don't check yourself into a geriatric ward just yet. The world is full of well-aged people who don't waste a moment dwelling on the ravages of time. Instead, they spend their advancing years joining political movements, writing novels, learning new crafts, and just generally continuing the trajectory of their lives. After all, there are advantages to living with a gently declining brain instead of one that's still immature, unpredictable, and changing fast.

Making the Most of an Aging Brain

Neuroscientists agree—although you can't guarantee that you'll dodge disease and bad luck, the best brain maintenance strategy is to compensate for age. In other words, add more to your brain than you lose. Here's how:

- **Practice lifelong learning.** By engaging your mind relentlessly, you can grow stronger synapses. Stronger synapses keep larger groups of neurons alive and well.

- **Remain engaged.** Seniors who remain engaged in social environments—for example, family, community groups, and (if you enjoy it) work—are likely to live longer and stay healthier. Once again, it's a simple case of stimulation keeping the brain alive.

- **Exercise.** Seniors who lead sedentary lives are easier targets for diseases like Alzheimer's. The exact cause isn't known. Exercise may help by reducing stress, staving off other health problems, or triggering synaptic growth.

- **Reduce stress.** Chronic stress damages the brain, leaving telltale scars in areas like the hippocampus. Avoid it, because you need to keep every neuron you can.

- **Treat other conditions.** Although there's no miracle supplement that can boost your brain power, neglecting your body can usher in other problems that will affect it. Some conditions that are correlated with poor brain performance, especially if they're untreated, include high blood pressure, depression, malnutrition, obesity, alcohol abuse, nicotine addiction, and diabetes.

- **Consider taking folic acid.** It's one of the few vitamins that was linked to reduced Alzheimer's risk in the infamous Nun Study (described next). It's not a slam dunk, but it's a compelling possibility.

Alzheimer's: The Nun Study

For most people, the most frightening part of brain aging is the risk of *Alzheimer's disease*, which chews through a healthy brain destroying memory, personality, and ultimately all cognitive function.

Although the cause of Alzheimer's disease isn't known, it's characterized by clumps of plaque that disturb the delicate balance of neurons in the brain. As the disease progresses, these tangles spread throughout the brain. First, they disrupt short-term memory. At this point, the disease is notoriously difficult to diagnose, because its effect resembles normal age-related memory decline. Next, Alzheimer's invades the **hippocampus**, the brain structure that forges long term memory (page 98), causing more serious memory problems. Finally, the plaque spreads through the brain and leaks into the upper levels of the cerebral cortex, where it can disrupt every aspect of the sufferer's personality. The brain that's left behind is dramatically shrunken and riddled with fluid-filled holes.

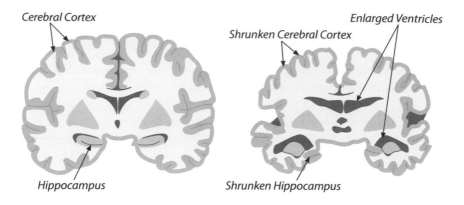

Cerebral Cortex

Enlarged Ventricles

Shrunken Cerebral Cortex

Hippocampus

Shrunken Hippocampus

Not long ago, neuroscientists considered Alzheimer's to be a natural part of aging. More recently, it's been reclassified as a degenerative disease, although it's still an open debate as to whether Alzheimer's is an inevitable consequence of aging. On average, roughly three percent of people over 65 suffer from Alzheimer's. At 85, this number rises to over 40 percent, and if you make it to 100 the odds are decidedly not in your favor. Although Alzheimer's isn't a genetic disease, if you have close relatives who suffer from Alzheimer's you're at a greater risk of developing it.

One of the most fascinating studies to tackle Alzheimer's is the Nun Study, which has followed the lives of 678 nuns in the U.S. The study's conclusions are particularly useful because its participants comprise a relatively consistent group. As you might expect from nuns, none are drug users, few drink alcohol, and most have similar life histories. These similarities minimize other factors that could confuse the results.

By far the most astounding finding from the Nun Study is that researchers can predict whether a nun will suffer from Alzheimer's by examining that nun's journal entries. The twist is that these journal entries were written by the nuns in their 20s, a full *60 years* before any of them would stare down the dreaded effects of the disease. Writers who wrote relatively plain journal entries, with few ideas and simple grammar, were far more likely to develop Alzheimer's 60 years later. By comparison, those who wrote grammatically complex, idea-rich entries stood a better chance of avoiding it altogether.

As compelling as these findings are, they don't quite solve the mystery of Alzheimer's. In fact, there are several possible explanations for these findings and no single obvious conclusion:

- **Learning prevents Alzheimer's.** In other words, the more you exercise your brain with reading, writing, and education, the better chance your brain will have of fending off degenerative diseases.

- **Learning compensates for Alzheimer's.** This is a similar argument, with an important twist. If you accept this argument, both the polished writers and the more prosaic ones stand an equal chance of getting Alzheimer's, but only the sharper-brained nuns have ways of compensating for it. This is a compelling argument, because there's no black-and-white test to diagnose Alzheimer's. When brains are dissected after death, they sometimes show the plaques and tangles of advanced Alzheimer's, even when the sufferer showed no decline in mental functioning.

- **People with Alzheimer-resistant brains are good learners.** If you accept this explanation, the idea-rich journal entries simply reflect something innately different about people who avoid Alzheimer's. If you don't share this quality, you can study your brain out without gaining any advantage.

Today, the debate is far from settled, and follow-up studies continue. However, the current evidence suggests that exercising your brain gives you the best shot at keeping it intact.

Nature vs. Nurture

Now that you've journeyed through the stages of life and seen how they shape your brain, there's only one question left. Namely, who should you thank for the undeniable wonderfulness of you?

For centuries, scientists have debated whether innate, inherited qualities (your *nature*) or personal experiences (the *nurture*) play the greater role in determining traits like personality and intelligence. The question is at least partly a matter of perspective. For example, if you compare the average person to a tree sloth, it's clear that genetic programming decides whether you're typing in cubicle or lounging in a subtropical tree. On the other hand, if you compare a modern bank teller to an eighth-century Tibetan monk, you might be inclined to think that environment has more than a passing influence on the way you spend your Monday mornings.

When scientists compare the influence of genes versus environment, they have a specific definition in mind. Essentially, the question scientists want to answer is this: if you gather together a large group of people, what accounts for the *variation* in their abilities? In other words, why can Joe outtalk, outcharm, and outromance Lenny, and why is Joan so much dafter than Sarah? When asked this way, the answer is easier to answer and no less important.

Heritability

To describe how strongly a specific characteristic depends on your genetic makeup, scientists use a measurement called *heritability*, which ranges from 0 to 1.

A heritability of 0 means the variance in a trait is entirely due to environmental factors. For example, language has a heritability of 0—if you speak English and your dentist speaks Hindi, it's because you were raised in different cultures.

A heritability of 1 means the variance in a trait is entirely up to the genes. For example, your blood type has a heritability of 1—it depends on your parents, not the unwritten rules of society.

Your height is obviously a bit more complex. The link between genes and height varies throughout your life but is strongest at adulthood, when the heritability sits at about 0.8. In other words, if you gather a group of people and measure their heights, about 80% of the variance can be explained by genetics. This is a high heritability, which makes a strong argument for all-in-the-family basketball picnics.

Heritability is a crude measure, because it reduces a complex interaction to a simple percentage. Many invisible and uncontrollable factors can influence the result. For example, if you calculate the heritability of height using last year's census numbers and historical records from the Great Irish Potato Famine, you'll probably find a much lower value. In this case, an environmental factor—the scarcity of starchy food—dominates over the usual expression of genetics.

As long as you understand that heritability is a comparative tool, not a definitive conclusion, you'll find that it's remarkably useful. It's particularly good at sizing up different traits to find out which one has the stronger genetic link. However, always remember that heritability applies to populations, not individual people. For example, if the heritability of IQ is 0.5 and your IQ tops the average score by 20 points, you can't thank your parents for their 10 point contribution (especially if they're statisticians). Life isn't that simple.

What's in a Name? Your Future Career

One example of questionable correlations is *nominative determinism*—the idea that a person's name influences their course in life. Researchers who explore this phenomenon find that there's a disproportionate number of people named Dennis in dentistry, an outsized amount of Geoffreys studying geology, and a surprising number of Florences moving to Florida. Furthermore, when strangers of the opposite sex meet, the ones with similar names (say, Eric and Erica) are more likely to start a romantic relationship.

Studies like these suggest that people prefer careers, homes, and romantic partners that subconsciously remind them of themselves. (Alternate explanations are possible. For example, people could be changing their names to fit their careers. Or, these patterns could be simple coincidences that would disappear in more expansive studies that examine more professions and more places.)

The magazine *New Scientist* occasionally invites readers to submit humorous examples of nominative determinism. Below are some amusing verified examples that can be found at *http://en.wikipedia.org/wiki/Nominative_determinism*:

- Dr. Richard (Dick) Chopp, a urologist known for vasectomies
- Marc Breedlove, a neuroscientist who's written several articles about sexuality
- William Shakespeare, a professor of English Literature at Brigham Young University.
- Cardinal Sin, the former Archbishop of Manila.

Technically speaking, these more colorful examples probably don't show nominative determinism at work. Instead, they're the product of a simple selection bias (page 158) that makes humorously appropriate names stand out from the crowd.

Family Studies

Researchers use a variety of statistical tricks to examine vast amounts of information and coax out heritability values. To collect the data they need, they run studies that compare family members and randomly picked strangers. Here are some examples:

- **Identical twins vs. strangers.** Identical twins have exactly the same genes (the nature part of the equation). If they're raised separately, they'll share none of the same environment (the nurture). Thus, if separated twins are more similar than a pair of randomly picked strangers, chalk it up to nature.

- **Identical twins vs. fraternal twins.** Fraternal twins are like any other pair of siblings—they share, on average, half their genes. (Older siblings know that younger siblings won't admit this fact, unless put in a headlock.) If identical twins are more similar than fraternal twins, it's due to nature.

- **Adoption studies.** Adopted children are genetic strangers—they share all of the family, with none of the genetic baggage. This allows many different types of comparisons. For example, if biological siblings are more similar than adopted siblings, it's another example of nature.

Although all of these studies are good for chewing up data and spitting out statistics, the most intriguing type is the first one, which compares identical twins who were separated at birth. Although it's not uncommon for researchers to find significant differences between separated twins, there's also an abundance of eerie similarities.

Reunited twins have discovered that they both enjoy cold cups of coffee, play on an invisible keyboard when deep in thought, wear the same tragically unfashionable hairstyle, and share the same taste for cinnamon. They've also discovered they share the same job, eccentric hobby, patterns of speech, life-changing events, and recurring dreams. Even without statistical number crunching, these stories hint that our genes have a powerful influence over our destinies. (Note for the skeptically inclined: the selection effect, described on page 158, encourages us to pay attention to striking similarities and ignore less interesting differences.)

 The idea of suddenly discovering that you have a genetic clone is irresistibly fascinating. Who wouldn't want to meet a person who was born with the same biological ingredients but took an alternate path through the accidents of life? For one telling of this story, you can read Elyse Schein and Paula Bernstein's *Identical Strangers: A Memoir of Twins Separated and Reunited* (Random House, 2007), which is the account of twins who discover each other (and the psychologist who separated them) at age 35.

Nature: Your Genes

For decades, the nurture side carried the day. They claimed that society was little more than a vast indoctrination factory that stamped morals and values onto each new member. In fact, leading psychologists argued that by carefully controlling environmental factors you could put any ordinary child on an irreversible path to success, crime, academic excellence, sexual promiscuousness, and so on. (You may remember how well this turned out for David Reimer, as described on page 211.)

This idea agreed with a lot of pleasant sentiments about life. For instance, it lent support to arguments that people have infinite flexibility to choose their path in life, and raising children is a science that can be fine-tuned as effectively as a soufflé recipe. However, in recent years the balance of research has tipped in the other direction. In other words, modern science suggests that many of the things that make you distinctively different from your peers start with the 23 pairs of chromosomes passed to you from your parents.

Here are some key findings:

- **IQ scores are highly heritable.** Different studies place the heritability of IQ between 0.4 and 0.8.

 IQ tests are always controversial—no one can agree about exactly what skills they measure, what they leave out, and whether it's even worth knowing in the first place. However, whatever IQ tests do measure, it's highly heritable.

- **Personality scores are highly heritable.** The big five personality dimensions (described in Chapter 9) also have a high heritability. So does religiosity and general happiness (in adulthood).

- **With most traits, heritability increases until adulthood.** For example, in early childhood height, weight, and IQ show a much lower heritability. The heritability increases through childhood and adolescence, and reaches its maximum value in adulthood. In other words, people are more similar to their parents as adults than as children.

 There are various arguments to explain why genetic links seem to get stronger with age. It's possible that until individuals are mature, the tests we have can't capture the relevant information. Or, it could be that a genetic advantage doesn't appear until development is finished (and a genetic disadvantage doesn't appear until development is held up somewhere along the way). Another possibility is that individuals compound minor genetic differences as they interact with their environments over time. For example, a person who begins with musical ability may attend concerts, pick up an instrument, and seek out training, all of which will strengthen that ability.

Nurture: Your Environment

There's a dirty secret hiding in even the very best heritability studies. Although researchers can pinpoint the importance of genes, they frequently fail when they try to explain the rest.

For example, you've already learned that about half the variance in IQ and personality stems from your genes. Naturally, this means the remaining bit is contributed by your environment. But when researchers go hunting for the exact environmental influence that's at work, they flounder helplessly through statistically irrelevant oceans of data. In fact, the list of factors they've tried and failed to match up is long and illuminating. Details about the parents—for example, their parenting style, the amount of time they spend with their children, their level of education, and so on—have an effect early on, but that effect diminishes with time until it's nearly undetectable. In fact, most studies find that *shared environmental influences* (influences that siblings have in common, such as the city they live in, the school they attend, the social standing of their family, and so on) have only the weakest effects. In adulthood, the correlation between the IQ and personality of an adopted child and that child's adopted parents is nearly 0.

So where are the missing environmental factors that are making all the difference? Here are some possibilities:

- **Peer groups.** In her highly controversial book *The Nurture Assumption* (Free Press, 1998), Judith Harris argues that siblings in the same family belong to different peer groups, and these peer groups exert more influence than any parents.

- **Epigenetic differences.** Recent research suggests that environmental factors can change how genes are expressed over time. Examples include minute differences in a baby's fetal environment or exposure to toxins like cigarette smoke throughout life. Epigenetic factors may explain the differences in identical twins—why they don't necessarily have the same fertility, why they don't enter menopause at the same age, and why one can suffer from a disorder like schizophrenia or manic depression while the other doesn't.

- **Individual variations.** Perhaps some of the environmental factors that researchers have measured are important, but only for small groups of people. For example, maybe parenting style makes a difference for some children who are more receptive to it, but not for others. This effect would still show up as a statistical correlation, but averaged out over a large population it would diminish and might be missed.

- **Too many similarities.** Perhaps parenting matters, but we're all doing a roughly equivalent job of it. Many of the large twin studies have collected data from a disproportionately white, middle class group of people. There's some evidence that more varied studies show more heritability for environmental factors. Of course, this suggests that parenting style matters, but not that much, and you're probably doing no better or no worse than your neighbor down the street.

 As you've learned throughout this book, the brain is addicted to cause-effect reasoning. For that reason, people often explain their personality based on environmental influences, even when the environmental influence is small. For example, in one case in a twin study, twins who had been raised apart had strikingly similar cleaning habits. (Both scrubbed and neatened obsessively.) When asked why, one twin explained he was emulating the relentless tidying of his tidy adopted parents, while the other said he was rebelling against the slovenly lifestyle of his adopted family.

The Interaction Between Genes and Environment

Many scientists argue that the entire debate is overly simplistic. After all, nature and nurture don't give separate contributions to a trait. Instead, they're locked in a complex relationship.

There are a few ways that environment can steal the limelight away from everyone's favorite double helix. First, extreme environments trump genetics. For example, undernourished people are likely to stay short and severely neglected children are likely to have low IQs, regardless of their genetic makeup. This is fairly obvious, but it's not always easy to spot the turning point where environment sneaks back into the picture. For example, one study found that a family's low economic standing could all but erase the genetic IQ link. In the poor families it considered, the heritability of IQ sank to near 0, while the influence of the home environment accounted for almost 60 percent of the variance.

Second, there are countless traits that have a significant genetic component, but only appear when you meet up with something specific in your environment. For example, you may have an exceptional ability that needs the right teachers or a potential brain disorder that needs the right traumatic trigger. This sort of relationship between genes and environment can quickly become quite complex. In fact, there are a number of ways that people seek out the right environment for the expression of their deep, dark, inner nature. Here are three ways that psychologists classify them:

- **Passive gene-environment correlations.** A child's home environment is based on the parents. The child's genetic makeup is also based on the parents. See the problem? Sociable parents are likely to have sociable children, and they're likely to make their children that much more sociable by exposing them to wild parties and big family events. Similarly, ambitious parents are likely to have ambitious children, and place a high value on homework.

- **Evocative gene-environment correlations.** An individual's personality also *evokes* certain reactions from other people. For example, a sociable child is more likely to become the star of the party and get invited to still more parties. Later in life, the same child might be picked for public speaking, placed in leadership positions, and asked to star in a series of skin cream commercials. All of these actions will force that child to develop stronger social skills.

- **Active gene-environment correlations.** Most obviously, people can choose to create an environment that suits their personality. They seek specific hobbies, careers, friends, and living arrangements that match their personalities. For example, a social person might head to the nightclub, embark on a traveling odyssey, or launch a career in sales.

Escaping Your Genetic Straitjacket

Many people don't react well when they learn that nature trumps nurture. They worry that biology is stealing away their free will. Oddly enough, when nurture wins the argument people spend relatively little time worrying that they'll be trapped by the accidents of their environment. Somehow, it's psychologically easier to accept that the world around you shapes your character than to admit that an invisible molecule of DNA is in the driver's seat.

Truthfully, there's no need to worry. Here are some points that should help you adopt a broader perspective:

- **Nature can't work without nurture.** Remember that even the most heritable trait needs the influence of your environment. It's up to you to search out the right environment to switch on your talents (and even better, richly reward them).

- **Heritability doesn't imply immutability.** In fact, it's perfectly possible to improve an IQ score, develop a new skill, or hone your personality. A highly heritable trait is simply one that, when considered over large groups of people with a wide variety of experience, is most clearly influenced by genes.

- **If you're a parent, don't feel too bad.** Although you may not have the ability to shape your child's character, you play a key role in supporting it. For example, you may not be able to make your child more musical or more athletic, but you can help him or her develop in either direction with piano lessons and soccer camp. Your contributions are the bricks that your child fits into a structure of his or her creation.

- **There's no shame in getting help.** Genes also control disorders, so teen problems like addiction, depression, and eating disorders may be already in the cards. For the best chance of resolving them, catch these problems early.

Index

Symbols

G

H

estrogen, 203
fight-or-flight, 136, 138
ghrelin, 35, 39
lack of sleep, 40
leptin, 39
oxytocin, 218–220
stress, 139
testosterone, 202–204
Hubbard, Edward, 105
humor, 149
hydrogenation, 34
hypnagogic imagery, 55
hypoglycemia, 29
hypothalamus, 21, 148
appetite, 38
biological clock, 46
controlling, 43
fear, 136
gender differences, 205
oxytocin, 218

I

identical twins
vs. fraternal twins, 242
vs. strangers, 241
immune system and stress, 139
improvising jazz musicians, 172
indexing memories with pegs, 116–117
information overload, 104
inkblot test, 85
insomnia, 53
insulin, 30, 31
IQ
after age 20, 236
environment, 244
heritability, 243
nature vs. nurture, 245
iron, 34

J

journals as memory aids, 118–119
journey method, 113
Jouvet, Michel, 59
judgment, 149

K

Kitaoka, Akiyoshi, 72
Kubrick, Stanley, 170

L

language, 14
after age 20, 236
children, 227
critical periods, 230
lateral thinking problems, 172
lateral thinking puzzle, 171
learning, 18, 28
Alzheimer's disease, 238
language, critical periods, 230
lifelong, 237
new skills (teenage years), 234
sleep, 59–60
new tasks, 60
tips, 120–122
toys, 231
LeDoux, Joseph, 14, 135
leptin, 39
lack of sleep, 40
set point theory, 41
sleep deprivation, 54
lexical hypothesis, 180
limbic system, 11
liver, 30, 31
loci, method of, 110–113
logic, 14
logical fallacies
argue from authority, 167
attacking arguer instead of argument, 167
circular logic, 169
criticizing consequences of a belief, 169
distorting opposing view, 167
presenting false choice, 168
shifting goalposts, 168
value-laden words, 166
long-term memory, 94, 97–106
Alzheimer's disease, 237
dendrites, 97
hippocampus, 98

Try the online edition free for 45 days

"The Missing Manual series is simply the most intelligent and usable series of guidebooks..."
—Kevin Kelly, co-founder of Wired

Your Brain
THE MISSING MANUAL.
How to get the most from your mind

POGUE PRESS™
O'REILLY®   Matthew MacDonald

Keeping your competitive edge depends on having access to information and learning tools on the latest creative design, photography and Web applications—but that's a lot easier said than done! How do you keep up in a way that works for you?

Safari offers a fully searchable library of highly acclaimed books and instructional videos on the applications you use every day. Best-selling authors offer thousands of how-tos, sharing what really works, and showing you how to avoid what doesn't. And with new content added as soon as it's published, you can be sure the information is the most current and relevant to the job at hand.

To try out Safari and the online edition of the above title FREE for 45 days, go to www.oreilly.com/go/safarienabled and enter the coupon code KXMSBWH

To see the complete Safari Library visit:
safari.oreilly.com

Safari
Books Online